Ecology of Urban Environments

Kirsten M. Parris
School of Ecosystem and Forest Sciences
The University of Melbourne
Melbourne, Australia

WILEY Blackwell

This edition first published 2016 © 2016 by John Wiley & Sons Ltd

Registered office: John Wiley & Sons, Ltd, The Atrium, Southern Gate, Chichester, West Sussex, PO19 8SQ, UK

Editorial offices: 9600 Garsington Road, Oxford, OX4 2DQ, UK
The Atrium, Southern Gate, Chichester, West Sussex, PO19 8SQ, UK
111 River Street, Hoboken, NJ 07030-5774, USA

For details of our global editorial offices, for customer services and for information about how to apply for permission to reuse the copyright material in this book please see our website at www.wiley.com/wiley-blackwell.

Library of Congress Cataloging-in-Publication Data applied for

ISBN: 9781444332643 (Hardback)
ISBN: 9781444332650 (Paperback)

A catalogue record for this book is available from the British Library.

Wiley also publishes its books in a variety of electronic formats. Some content that appears in print may not be available in electronic books.

Set in 9/12pt, MeridienLTStd by SPi Global, Chennai, India.

1 2016

For Mick and Owen, my biggest supporters

and

In memory of Joanne Ainley, urban ecologist

Contents

Foreword

In the first quarter of the 21st century, we are living in a unique time in history – we are witnessing the move to the Anthropocene, a geological period in which humans have become a major driver of planetary processes. This time has also been termed the "Great Acceleration", a period of rapid increase in a large array of human activities with correspondingly large impacts on many environmental and biological processes. One of the clearest indicators of the changes underway is the emergence of the city as the main human habitat. For most of human history, the majority of the human population lived in rural or extensive landscapes, with only a few settlements that could be called "cities". Rapid human-population increase and industrialization from the 1800s onwards has been paralleled by a dramatic increase in the number and size of cities and an ever-increasing proportion of the human population living in urban areas. For the first time in history, humans are predominantly an urban species.

Cities are therefore extremely important environments from a human perspective: they are where most people now live. They form the centres of economic and cultural activity and are as diverse as the economies and cultures that created them. Cities come in all shapes and sizes, some developing from early hubs of trade and navigation and some springing up in entirely new locations. Early cities tended to be compact and geared for travel by foot or horse. However, cheap cars and mass transit have released cities from their earlier spatial constraints and many now sprawl extensively in all directions. Planning and management of urban form and function have become increasingly important endeavours as cities evolve, grow and require more efficient and effective private buildings, public spaces and essential services.

Cities are not only home to humans, however. Most cities are mosaics of built infrastructure and open space – parks, gardens, waterways and remnants of the nature that was present prior to the city's construction. These spaces are inhabited by a wide range of species, some of which thrive in the urban environment and some of which struggle. The mix of species present can include many species native to the region and many that have been introduced or have adapted themselves to the urban environment. Just as the city provides a focus for human creativity, it can also act as a place where new biologies play out – new combinations of species, species doing new things, and species interacting in novel ways with humans and their built environment.

Given the increasing importance of the city environment and the richness and fascination of the biological systems that develop in cities, it might seem odd that the field of urban ecology has only blossomed quite recently. Indeed, a few decades ago, it would appear that cities were not regarded as places where ecologists would want to work – seeking out instead the remote, apparently untouched areas where the ecology was "intact". Today, cities are seen more as one end of a spectrum of humanized landscapes and are increasingly the subject of research into their ecological function. How do all the species found in cities persist and thrive? How do ecological communities develop within the altered environments found in cities? How does the urban ecosystem "work" with respect to flows of water, nutrients and energy? How do humans relate to, modify, and live in these environments? And, finally, are there better ways to plan and manage cities and their components that lead to greater liveability for humans and diverse biological communities alike?

This is the stuff of urban ecology, a growing field that seeks to understand how cities work in terms of their ecology and in relation to both their human and non-human inhabitants. This book is, a very timely contribution that provides an accessible yet fascinating synthesis of the ecology of urban environments. Within a strong framework of existing ecological theory, it explores how the construction and expansion of cities influence the characteristics of urban environments and the dynamics of populations, communities and ecosystems. It considers the ecology of human populations in cities, and presents a compelling case for conserving biodiversity and maintaining ecosystem services in urban landscapes. Overall, it seeks to help us better understand, plan and manage our primary habitat – a task that gets more pressing and important by the minute, both for ourselves and for the other species with which we share our cities.

Richard J. Hobbs
University of Western Australia

Preface

A long time ago came a man on a track
Walking thirty miles with a sack on his back
And he put down his load where he thought it was the best
Made a home in the wilderness

He built a cabin and a winter store
And he ploughed up the ground by the cold lake shore
The other travellers came walking down the track
And they never went further, no, they never went back

Then came the churches, then came the schools
Then came the lawyers, then came the rules
Then came the trains and the trucks with their loads
And the dirty old track was the Telegraph Road

Then came the mines, then came the ore
Then there was the hard times, then there was a war
Telegraph sang a song about the world outside
Telegraph Road got so deep and so wide
Like a rolling river

And my radio says tonight it's gonna freeze
People driving home from the factories
There's six lanes of traffic
Three lanes moving slow

I used to like to go to work but they shut it down
I got a right to go to work but there's no work here to be found
Yes, and they say we're gonna have to pay what's owed
We're gonna have to reap from some seed that's been sowed

And the birds up on the wires and the telegraph poles
They can always fly away from this rain and this cold
You can hear them singing out their telegraph code
All the way down the Telegraph Road

Mark Knopfler,
Telegraph Road

Acknowledgements

I have many people to thank for their encouragement, support and enthusiasm, without which this book would have remained unwritten. Alan Crowden prompted me to consider writing a text book on urban ecology; Mark Burgman convinced me that it was a good idea. Alan facilitated the book's publication with Wiley-Blackwell, and both he and Mark have been important supporters and mentors from its inception to completion.

I am grateful to many colleagues for their interest in this project and for helpful discussions about the ecology of urban environments, including Sarah Bekessy, Stefano Canessa, Jan Carey, Yung En Chee, Martin Cox, Danielle Dagenais, Jane Elith, Carolyn Enquist, Brian Enquist, Fiona Fidler, Tim Fletcher, Georgia Garrard, Leah Gerber, Gurutzeta Guillera-Arroita, Amy Hahs, Josh Hale, Andrew Hamer, Geoff Heard, Samantha Imberger, Claire Keely, Jesse Kurylo, José Lahoz-Monfort, Pia Lentini, Steve Livesley, Adrian Marshall, Mark McDonnell, Larry Meyer, Joslin Moore, Alejandra Morán-Ordóñez, Raoul Mulder, Emily Nicholson, Cathy Oke, Joanne Potts, Hugh Possingham, Dominique Potvin, Peter Rayner, Tracey Regan, John Sabo, Caragh Threlfall, Reid Tingley, Rodney van der Ree, Peter Vesk, Chris Walsh, Andrea White, Nick Williams and Brendan Wintle.

I thank all my family, friends and colleagues for their support throughout the process of writing this book. Particular thanks to Michael McCarthy, Owen Parris, Ann Parris, John Gault, Bronwyn Parris, Monica Parris, Bridget Parris, Susan McCarthy, David McCarthy, Margery Priestley, Liz McCarthy, Tom McCarthy, Kirsty McCarthy, Sarah Bekessy, Michael Bode, Gerd Bossinger, Lyndal Borrell, Janine Campbell, Jan Carey, Jane Catford, Tasneem Chopra, Glenice Cook, Martin Cox, Kylie Crabbe, Karen Day, Jane Elith, Louisa Flander, Jane Furphy, Georgia Garrard, Cindy Hauser, Colin Hunter, Helen Kronberger, Rachel Kronberger, Min Laught, Sue Lee, Prema Lucas, Pavlina McMaster, Ruth Millard, John Moorey, Anne Macdonald, Meg Moorhouse, Sarah Niblock, Lisa Palmer, James Panichi, Rebecca Paton, Joanne Potts, Tracey Regan, Di Sandars, Anna Shanahan, Peter Vesk, Graham Vincent, Terry Walsh, Andrea White, Brendan Wintle and Ian Woodrow.

The Australian Research Council, the Faculty of Science at The University of Melbourne, the Australian Research Centre for Urban Ecology, and the NESP Clean Air and Urban Landscapes Hub have provided valuable support for my research on urban ecology. I thank my intrepid and enthusiastic Research Assistant, Larry Meyer, who has helped me with many aspects of this project. I am very grateful to Mark Burgman, Michael McCarthy, Caragh Threlfall,

Chris Walsh and the 2014 Graduate Seminar: Environmental Science class at The University of Melbourne for reading and providing insightful comments on the draft manuscript. Lastly, I thank Ward Cooper, Delia Sandford, Kelvin Matthews, Emma Strickland and David McDade at Wiley-Blackwell and Kiruthika Balasubramanian at SPi Global for their assistance and patience as this book became a reality.

Telegraph Road
Words and Music by Mark Knopfler

CHAPTER 1

Introduction

1.1 Setting the scene

Reading this book, there is a good chance that you live in an urban environment – a town or a city. And if you look out of your window or door, you might see buildings, roads, cars, fences and street lights, as well as people, cats, dogs, trees or flowers. You might hear a train rumbling, a jackhammer hammering, a violin playing, children laughing or birds singing. You might smell diesel exhaust from a passing truck, risotto cooking at a nearby restaurant, newly-mown grass from the park across the road, or the stench of a rubbish heap or an open drain. These are the contrasts of life in the city, where the best and worst of human existence can be found, and where habitats constructed for people can complement or obliterate the habitats of other species. Ecologists strive to understand the processes of and patterns in the natural world. Until recently, many ecologists practised their science in places far from cities, considering human activity to be a disruption – rather than a part – of nature. But ecological principles apply in urban environments too, and the separation of humans from the rest of nature occurs to our detriment. Urban ecology is a relevant and valuable discipline in the highly-urbanized world of the 21st century.

1.2 What is urban ecology?

As a natural science within the broader discipline of biology, ecology is the study of the distribution, abundance and behaviour of organisms, their interactions with each other and with their environment. Ecology traverses many scales, from within individual organisms to whole individuals, populations, communities and ecosystems. Organisms are living things, such as bacteria, fungi, plants and animals. Human animals (people) have not generally been studied alongside other organisms as part of ecology (but see human behavioural ecology: Winterhalder and Smith 2000; Borgerhoff Mulder and Schacht 2012). This is the first point of difference between urban ecology and other ecological disciplines; the second is its focus on urban environments, which can be considered as habitats designed by people for people.

In this book, I define urban ecology as the ecology of all organisms – including humans – in urban environments, as well as environments that are impacted by

Ecology of Urban Environments, First Edition. Kirsten M. Parris.

the construction, expansion and operation of cities, such as forested watersheds (catchments) that supply drinking water to urban populations. Urban ecology includes people because the presence, population dynamics and behaviour of people, and the environmental changes that occur when they construct towns and cities, are central to our understanding of how urban systems function. Urban ecology has a different meaning in the social sciences, where it describes an approach to urban sociology that uses ecological theory to understand the structure and function of cities (e.g., Park and Burgess 1967). Some authors also use the term urban ecology to describe an interdisciplinary field that brings together the natural sciences, social sciences and humanities (e.g., Dooling et al. 2007; see Chapter 8 for further discussion of this point). However, the motivation for and focus of this book are strongly grounded in the natural science of ecology. Ecology has much to offer the study of cities and towns, and this book provides a conceptual synthesis of the extensive but often disparate urban-ecological literature. In combination with other disciplines in the natural sciences, social sciences and humanities, an improved understanding of urban ecology will make a vital contribution to improved urban planning, design and management, for the benefit of all species that live in cities.

Urban ecology is a relatively young discipline and there has been some debate about what it should encompass and how the term "urban" should be defined (e.g., Collins et al. 2000; McIntyre et al. 2000; Pickett et al. 2001). For example, should we recognize an urban area by the number or density of people living there, by certain characteristic landscape patterns, by the density of features such as buildings and roads, or a combination of these things (McIntyre et al. 2000; Luck and Wu 2002; Hahs and McDonnell 2006)? Is there a single definition of urban that everybody should use, or are there a number of acceptable definitions that are suitable for different research questions? Wittig (2009) supports a very narrow definition of the term urban, as inner-city neighbourhoods dominated by concrete, asphalt and buildings, with no original vegetation remaining. This excludes other parts of cities, such as streams, private gardens and areas of remnant vegetation. It also excludes environments outside towns and cities that are nonetheless impacted by them. Pursuit of one definition of "urban" to be used in all urban-ecological studies may not be very useful, as definitions are likely to change with the scale of a study and the questions being asked. What is urban for a stream or an owl may differ from what is urban for a person, a beetle or a fungus. However, it is important that the definition is both clear and quantitative to allow the methods of a study to be replicated, and to assist comparison between studies and formal meta-analysis (McIntyre et al. 2000).

1.3 Why is urban ecology interesting?

Urban ecology is interesting for at least five reasons: (i) urban environments are extensive and growing; (ii) their ecology is inherently interesting; (iii) they are ideal for testing and developing ecological theory; (iv) the nature of urban environments affects the health and wellbeing of their human inhabitants and (v) they are important for conserving biological diversity. An improved understanding of urban ecology will not only advance the discipline of ecology as a whole, it will help us to save species from extinction, maintain ecosystem functions and services, and improve human health and wellbeing. Particularly in these times of rapid human-population growth and urbanization, a better understanding of urban environments will help us to create more liveable cities that provide high-quality habitat for humans and non-humans alike. I address each of these points in more detail below.

1.3.1 Urban environments are extensive and growing

For the first time in history, more than half the world's human population lives in urban areas. The number of people living in cities has risen dramatically since the industrial revolution, as opportunities for employment have expanded in urban areas and the demand for agricultural labour has declined with increasing mechanization. The United Nations Population Fund (UNFPA) estimates that the world's current urban population of 3.9 billion people will expand to 4.9 billion by 2030 and 6.4 billion by 2050 (Figure 1.1a), compared to an urban population of just 220 million at the beginning of the 20th century (UNFPA 2007; UN 2014). This equates to a 22-fold increase in only 130 years. Urban areas in the developed world will grow slightly, while much of the expected increase in the number of people living in towns and cities will occur in developing countries in Africa, Asia, Latin America and the Caribbean (Fig 1.1b; UNFPA 2007). The social and environmental implications of the shift to urban living are profound, but they also vary dramatically between regions.

Urban expansion in developed countries such as Australia and the USA is typically accommodated through the construction of houses on individual blocks of land on the outskirts of towns and cities (Figure 1.2). Most houses are inhabited by a single family, and have electricity, potable tap water, one or more bathrooms connected to a closed sewage system, a telephone and a sealed road at their front door. Some houses have swimming pools; many have air-conditioners. Relatively large areas of land accommodate only a few people, and the resulting expansion of cities across the landscape is known as urban sprawl (Soule 2006). In contrast, many people moving to urban areas in sub-Saharan Africa, Latin America, India and China are accommodated in informal settlements (also known as slums or shanty towns) within or on the edges of cities (UNFPA 2007). These are characterized by a high density of people living in makeshift dwellings with poor sanitation, little or no access to clean drinking water, and uncertain tenure (Figure 1.3). Hundreds of people may share a single bathroom; water used for drinking can be contaminated with human waste; dwellings often have no electricity or ventilation; and there are no paved roads or facilities for waste disposal (Geyer et al. 2005; UNFPA 2007). Informal settlements are frequently built in areas subject to natural disasters, such as floods and landslides, and because the people who live there have no contractual right to do so, their dwellings can be demolished at short notice (Hardoy and

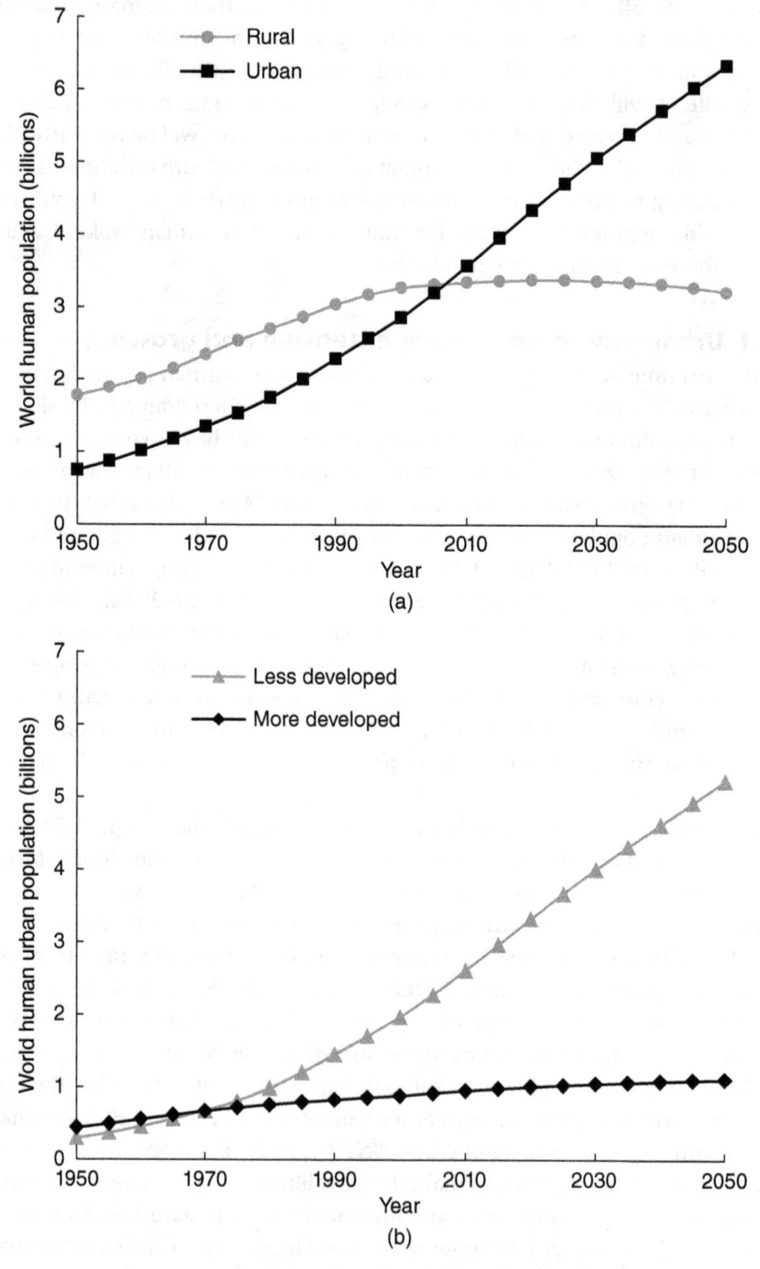

Figure 1.1 (a) World population of humans in urban and rural areas and (b) the urban population of humans in less developed and more developed regions of the world, 1950–2050. Data from United Nations, Department of Economic and Social Affairs, Population Division (2014).

Figure 1.2 An ordered suburb in Las Vegas, Nevada, USA. Picture has been straightened, cropped and converted to black and white. Photograph by ulybug. Used under CC-BY-2.0. https://creativecommons.org/licenses/by/2.0/.

Figure 1.3 Houses in the Kibera Slum, Nairobi, Kenya. Picture has been cropped and converted to black and white. Photograph by Colin Crowley. Used under CC-BY-2.0 https://creativecommons.org/licenses/by/2.0/.

Satterthwaite 1989; Tibaijuka 2005; Padhi 2007). An estimated 1 billion people, or one-sixth of the world's population, were living in informal settlements in 2005 (UN-Habitat 2006).

Other styles of urban development fall between these two extremes. Medium- and high-density townhouses and apartments are features of urban living in many parts of the world, providing a high standard of living with modern facilities and infrastructure but occupying less space than detached houses. Each type of urban expansion affects the biological diversity and ecosystem function of the newly-urbanized areas, and other associated habitats, in very different ways (Liu et al. 2003). For example, the materials and energy used to construct large, low-density houses on the outskirts of an Australian city and to maintain the lifestyle of their inhabitants are many times greater than those used by the inhabitants of a shantytown in South Africa or Bangladesh (Wackernagel and Rees 1996; McGranahan and Satterthwaite 2003). The health and wellbeing of the people living in each type of urban neighbourhood also varies dramatically (see later in this chapter for further discussion of this point). Therefore, we cannot think of urbanization (the construction of towns and cities) or urban expansion (an increase in the human population of cities) as uniform processes. In the coming decades, urban expansion in the developing world will present enormous ecological and social challenges. I argue that these challenges will be better met with an improved understanding of urban ecology.

1.3.2 Urban environments have inherent ecological interest

Urban environments are of intrinsic ecological interest, partly because they can be so different from the habitats they replace. How do ecosystems, communities, species and populations adapt to the dramatic changes associated with the conversion of wild or agricultural land into habitat for people? Which species and communities thrive and which suffer? Do novel biological communities arise when native species are lost and exotic species invade? If so, are these communities functionally similar to the ones they replace, even though they are compositionally different? Can urban areas function as urban ecosystems? Is there a particular level of urbanization at which ecosystem function breaks down? What are the relationships between human preferences and actions and the conservation of biological diversity in cities? And how is human health influenced by air pollution, tree cover or access to open space? We know the answers to some of these questions for some parts of the world, but there are many more relationships between non-human organisms, humans and their environment in cities to be further explored.

1.3.3 Urban environments are ideal for testing and developing ecological theory

Much ecological theory has been developed to explain the distribution, diversity, behaviour and interactions of organisms in relatively pristine habitats away from human disturbance. Examples include the theory of the ecological niche

(Hutchinson 1957), interspecific competition (Tansley 1917; Connell 1961), optimal foraging theory (Charnov 1976), predator-prey relations (Volterra 1926; Lotka 1932), the equilibrium theory of island biogeography (MacArthur and Wilson 1963, 1967), metapopulation theory and patch dynamics (Levins 1969; Pickett and White 1985), food webs (Hairston et al. 1960; Murdoch 1966), metacommunity theory (Gilpin and Hanski 1991; Leibold et al. 2004) and the neutral theory of biodiversity and biogeography (Hubbell 2001). Behavioural theories such as game theory (Maynard Smith and Price 1973) and those pertaining to animal communication, mate choice and sexual selection (e.g., Zahavi 1975; Marten and Marler 1977; Wells 1977; Kirkpatrick and Ryan 1991) have also been developed largely without reference to the behaviour of animals in urban settings.

As argued by Collins et al. (2000), any worthwhile ecological theory should apply to urban as well as rural or wild environments. Theories that have been put to the urban test, such as optimal foraging theory (Shochat et al. 2004), niche theory (Parris and Hazell 2005), the intermediate disturbance hypothesis (Blair and Launer 1997), metacommunity theory (Parris 2006), diversity–productivity relationships (Shochat et al. 2006) and food webs/trophic dynamics (Faeth et al. 2005), have all fared well. This suggests that much – if not all – existing ecological theory is applicable to urban areas. The dynamic nature of urban environments may also encourage the development of new ecological theory, as well as new ways to integrate ecological, social and economic theories to understand better the ecology of urban systems.

1.3.4 The nature of urban environments affects human health and wellbeing

The nature of our surroundings affects human health and wellbeing in obvious and subtle ways. In urban areas, the starkest contrast in health and wellbeing is between people living in secure, well-constructed housing and those who are homeless or living in informal settlements. Inadequate sanitation, limited access to clean drinking water and poor protection from extremes of weather dramatically increase the risk of disease in slum communities, while a lack of privacy and security exposes women to violence and sexually-transmitted infections such as HIV-AIDS (Amuyunzu-Nyamongo et al. 2007; UNFPA 2007). But characteristics of the urban environment also affect the health of the adequately-housed urban dweller. Access to green nature and open space in cities provides opportunities to exercise and improves mental health (Giles-Corti et al. 2005; Gidlöf-Gunnarsson and Öhrström 2007). Recent research has shown that urban sprawl is correlated with increased rates of obesity and an increased risk of traffic and pedestrian fatalities (Ewing et al. 2003, 2006, 2016; Smith et al. 2008; Mackenbach et al. 2014). Sprawling neighbourhoods often have few footpaths, and facilities such as schools and shops are separated from residential areas. As a consequence, residents are more likely to drive their cars than to walk or cycle (Ewing et al. 2016).

At its best, urban living provides opportunities for social interaction and a sense of community (social capital), which are both important for human wellbeing. However, social capital can be eroded in cities with high crime rates, overcrowded living conditions, or conversely, when sprawling development leads to social isolation (Leyden 2003). High social disorder in urban areas is correlated with an increased risk of clinical depression among residents (Kim 2008). An intriguing area for further research is the relationship between biodiversity and human health in cities; recent studies have found that the psychological benefits of parklands and other green space for human visitors increase with increasing biodiversity (Fuller et al. 2007; Carrus et al. 2015).

1.3.5 Urban environments are important for conserving biological diversity

In the past, many towns and cities were established next to rivers, estuaries or sheltered harbours, which provided both attractive surroundings and opportunities to transport goods and people. Such sites also tended to be high in biological diversity (biodiversity) because of their high productivity, relatively mild climate, and position at the confluence of terrestrial, riverine and marine habitats (Luck 2007). The correlation between human population density and biodiversity continues today, with species-rich areas still being preferentially settled by people (Cincotta et al. 2000; Luck et al. 2004; but see Box 2.1). For example, urban development in Australia is proceeding along the coastal fringe of the continent, where rainfall, primary productivity and biodiversity are high. Because of the dramatic environmental changes it entails, urbanization often creates a conflict between the needs of humans and the needs of other species.

Throughout human history, urbanization has probably caused the local extinction of thousands of species. McDonald et al. (2008) estimated that 420 species (8%) of those included on the IUCN Red List are threatened by urbanization. Currently, 11 of the world's 825 ecoregions have over half their area urbanized (Figure 1.4) and 29 ecoregions have over one-third of their area urbanized (McDonald et al. 2008). These 29 ecoregions are home to 3056 species including 213 endemic terrestrial vertebrate species, 89 of which are included on the IUCN Red List. Particular functional groups are more likely to be lost from urban areas, such as ground-dwelling arthropods, insectivorous birds, large-bodied carnivores and ground-dwelling vertebrates that are vulnerable to introduced predators (Sewell and Catterall 1998; van der Ree and McCarthy 2005; Bond et al. 2006; Riley 2006). Short, shade-loving plants that require high levels of soil moisture are more likely to be lost from urban areas in Britain, while tall plants that favour open, dry habitats thrive (Thompson and McCarthy 2008; Duncan et al. 2011).

Even as urbanization leads to the loss of native species, many non-native plants and animals are introduced to towns and cities – either inadvertently or deliberately as pets and garden plants (McKinney 2002, 2008; Tait et al. 2005). A few species show a strong positive response to the resources provided

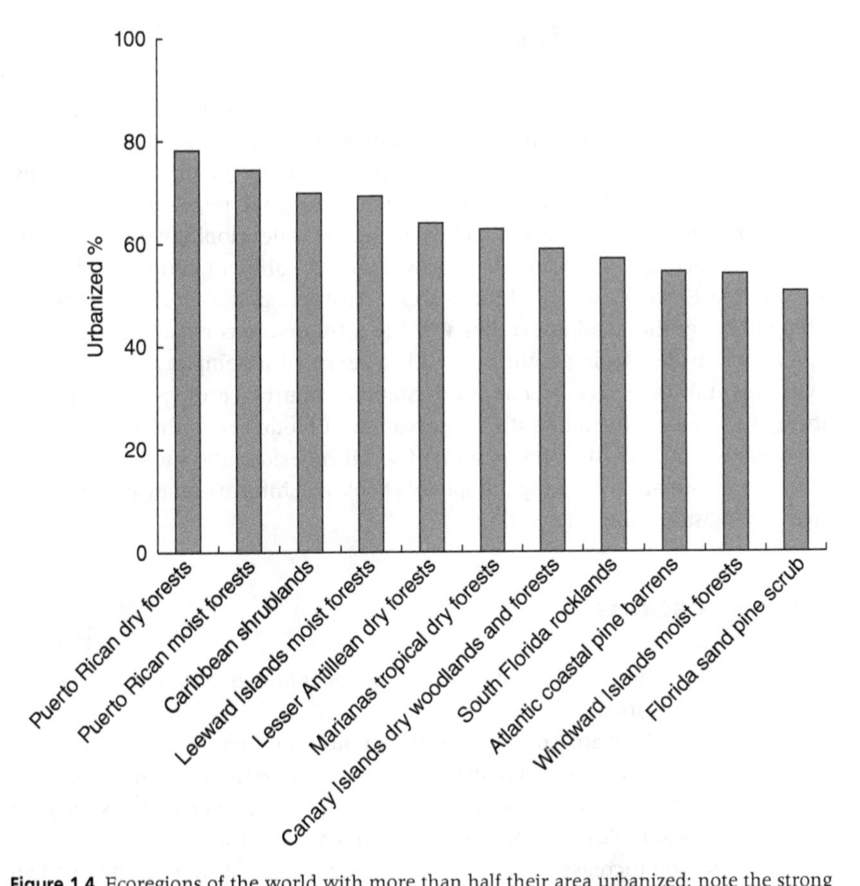

Figure 1.4 Ecoregions of the world with more than half their area urbanized; note the strong bias towards islands and coastal areas. Endemic species in these regions are threatened by continued urban expansion. Data from McDonald et al. (2008), supplementary information.

by humans in cities, where they obtain very high population densities. These are known variously as commensal species, synanthropes or urban adapters (see Chapter 7). However, this should not lead us to dismiss urban areas as insignificant for biological conservation. Many threatened, native species and ecological communities persist in or on the margins of towns and cities (e.g., Williams et al. 2005; Marchetti et al. 2006; Ives et al. 2016), particularly those that have developed slowly (Vähä-Piikkiö et al. 2004). Continuing urban expansion is likely to jeopardize the persistence of these species and communities, plus others that are currently at a comfortable distance from urban centres (McDonald et al. 2008), unless we change the way we construct and manage cities. A combination of ecological knowledge and careful urban planning both inside and outside nature reserves is required to minimize the loss of biodiversity from existing and future urban areas (Luck et al. 2004).

1.4 The aims of this book

This book has two aims. The first is to provide an accessible introduction to urban ecology, by synthesizing existing knowledge and using established ecological theory to identify generalities in the complexity of urban ecosystems. The second is to make urban ecology interesting and relevant to students, researchers and policy-makers in the developed and developing world. To date, much urban-ecological research has focussed on the affluent countries of North America and Europe and, to a lesser extent, Australia and New Zealand. But as outlined above, the coming decades will see a tremendous expansion of urban populations in developing countries, with a range of accompanying social and environmental challenges. A better understanding of urban ecology is vital for the future of the Earth, including the conservation of biodiversity, the maintenance of ecosystem function, the preservation of social cohesion, and the improvement of human health and wellbeing. I hope this book will inform and inspire budding urban ecologists around the world.

Study questions

1. How would you define the term "urban"? Consider both qualitative descriptions and quantitative metrics.
2. Which aspects of urban ecology interest you most, and why?
3. Describe the contrasts between formal and informal settlements in cities.
4. Are urban areas important for the conservation of biodiversity? Justify your answer with examples from at least three cities around the world.
5. How would you increase contact between people and nature in a city or town that you know? Consider a number of different strategies.

References

Amuyunzu-Nyamongo, M., Okeng'O, L., Wagura, A. *et al.* (2007) Putting on a brave face: The experiences of women living with HIV and AIDS in informal settlements of Nairobi, Kenya. *AIDS Care*, **19**, 25–34.

Blair, R.B. and Launer, A.E. (1997) Butterfly diversity and human land use: species assemblages along an urban gradient. *Biological Conservation*, **80**, 113–125.

Bond, J.E., Beamer, D.A., Lamb, T. *et al.* (2006) Combining genetic and geospatial analyses to infer population extinction in mygalomorph spiders endemic to the Los Angeles region. *Animal Conservation*, **9**, 145–157.

Borgerhoff Mulder, M. and Schacht, R. (2012) Human Behavioural Ecology. *eLS, Wiley*. doi: 10.1002/9780470015902.a0003671.pub2

Carrus, G., Scopelliti, M., Lafortezza, R. *et al.* (2015) Go greener, feel better? The positive effects of biodiversity on the well-being of individuals visiting urban and peri-urban greens areas. *Landscape and Urban Planning*, **134**, 221–228.

Charnov, E.L. (1976) Optimal foraging: the marginal value theorem. *Theoretical Population Biology*, **9**, 129–136.

Cincotta, R.P., Wisnewski, J., and Engelman, R. (2000) Human population in the biodiversity hotspots. *Nature*, **404**, 990–992.

Collins, J.P., Kinzig, A., Grimm, N.B. *et al.* (2000) A new urban ecology. *American Scientist*, **88**, 416–425.

Connell, J.H. (1961) The influence of interspecific competition and others factors on the distribution of the barnacle *Chthamalus stellatus*. *Ecology*, **42**, 710–723.

Dooling, S., Graybill, J., and Greve, A. (2007) Response to Young and Wolf: goal attainment in urban ecology research. *Urban Ecosystems*, **10**, 339–347.

Duncan, R.P., Clemants, S.E., Corlett, R.T. *et al.* (2011) Plant traits and extinction in urban areas: a meta-analysis of 11 cities. *Global Ecology and Biogeography*, **20**, 509–519.

Ewing, R., Brownson, R.C., and Berrigan, D. (2006) Relationship between urban sprawl and weight of United States youth. *American Journal of Preventive Medicine*, **31**, 464–474.

Ewing, R., Hamidi, S., and Grace, J.B. (2016) Urban sprawl as a risk factor in motor vehicle crashes. *Urban Studies*, **53**, 247–266.

Ewing, R., Schieber, R.A., and Zegeer, C.V. (2003) Urban sprawl as a risk factor in motor vehicle occupant and pedestrian fatalities. *American Journal of Public Health*, **9**, 1541–1545.

Faeth, S.H., Warren, P.S., Shochat, E. *et al.* (2005) Trophic dynamics in urban communities. *BioScience*, **55**, 399–407.

Fuller, R.A., Irvine, K.N., Devine-Wright, P. *et al.* (2007) Psychological benefits of greenspace increase with biodiversity. *Biology Letters*, **3**, 390–394.

Geyer, N., Mmuwe-Hlahane, S., Shongwe-Magongo, R.G. *et al.* (2005) Contributing to the ICNP: validating the term 'informal settlement'. *International Nursing Review*, **52**, 286–293.

Gidlöf-Gunnarsson, A. and Öhrström, E. (2007) Noise and well-being in urban residential environments: The potential role of perceived availability to nearby green areas. *Landscape and Urban Planning*, **83**, 115–126.

Giles-Corti, B., Broomhall, M., Knuiman, M. *et al.* (2005) Increasing walking – How important is distance to, attractiveness, and size of public open space? *American Journal of Preventive Medicine*, **28**, 169–176.

Gilpin, M.E. and I.A. Hanski (eds). 1991. *Metapopulation Dynamics: Empirical and Theoretical Investigations*. Academic Press, London.

Hahs, A.K. and McDonnell, M.J. (2006) Selecting independent measures to quantify Melbourne's urban-rural gradient. *Landscape and Urban Planning*, **78**, 435–448.

Hairston, N.G., Smith, F.E., and Slobodkin, L.B. (1960) Community structure, population control, and competition. *American Naturalist*, **44**, 421–425.

Hardoy, E. and Satterthwaite, D. (1989) *Squatter Citizen. Life in the Urban Third World*, Earthscan Publications, London.

Hubbell, S.P. (2001) *The Unified Neutral Theory of Biodiversity and Biogeography*, Princeton University Press, Princeton.

Hutchinson, G.E. (1957) Concluding remarks, in *Cold Spring Harbour Symposium on Quantitative Biology 22*, Cold Spring Harbor, New York, pp. 415–427.

Ives, C.D., Lentini, P.E., Threlfall, C.G. *et al.* (2016) Cities are hotspots for threatened species. *Global Ecology and Biogeography* **25**, 117–126.

Kim, D. (2008) Blues from the neighborhood? Neighborhood characteristics and depression. *Epidemiologic Reviews* **30**, 101–117. doi: 10.1093/epirev/mxn009

Kirkpatrick, M. and Ryan, M.J. (1991) The paradox of the lek and the evolution of mating preferences. *Nature*, **350**, 33–38.

Leibold, M.A., Holyoak, M., Mouquet, N. *et al.* (2004) The metacommunity concept: a framework for multi-scale community ecology. *Ecology Letters*, **7**, 601–613.

Levins, R. (1969) Some demographic and genetic consequences of environmental heterogeneity for biological control. *Bulletin of the Entomological Society of America*, **15**, 237–240.

Leyden, K.M. (2003) Social capital and the built environment: the importance of walkable neighborhoods. *American Journal of Public Health*, **93**, 1546–1551.

Liu, J.G., Daily, G.C., Ehrlich, P.R. *et al.* (2003) Effects of household dynamics on resource consumption and biodiversity. *Nature*, **421**, 530–533.

Lotka, A.J. (1932) The growth of mixed populations: two species competing for a common food supply. *Journal of the Washington Academy of Sciences*, **22**, 461–469.

Luck, G.W. (2007) The relationships between net primary productivity, human population density and species conservation. *Journal of Biogeography*, **34**, 201–212.

Luck, G.W., Ricketts, T.H., Daily, G.C. *et al.* (2004) Alleviating spatial conflict between people and biodiversity. *Proceedings of the National Academy of Sciences of the United States of America*, **101**, 182–186.

Luck, M. and Wu, J. (2002) A gradient analysis of the landscape pattern of urbanization in the Phoenix metropolitan area of USA. *Landscape Ecology*, **17**, 327–339.

MacArthur, R.H. and Wilson, E.O. (1963) An equilibrium theory of insular biogeography. *Evolution*, **17**, 373–387.

MacArthur, R.H. and Wilson, E.O. (1967) *The Theory of Island Biogeography*, Princeton University Press, Princeton.

Mackerbach, J.D., Rutter, H., Compernolle, S. *et al.* (2014) Obesogenic environments: a systematic review of the association between the physical environment and adult weight status, the SPOTLIGHT project. *BMC Public Health*, **14**, 1–15.

Marchetti, M.P., Lockwood, J.L., and Light, T. (2006) Effects of urbanization on California's fish diversity: Differentiation, homogenization and the influence of spatial scale. *Biological Conservation*, **127**, 310–318.

Marten, K. and Marler, P. (1977) Sound transmission and its significance for animal vocalization I. Temperate habitats. *Behavioral Ecology and Sociobiology*, **2**, 271–290.

Maynard Smith, J. and Price, G.R. (1973) The logic of animal conflict. *Nature*, **246**, 15–18.

McDonald, R.I., Kareiva, P., and Forman, R.T.T. (2008) The implications of current and future urbanization for global protected areas and biodiversity conservation. *Biological Conservation*, **141**, 1695–1703.

McDonnell, M.J. and Pickett, S.T.A. (1990) The study of ecosystem structure and function along urban-rural gradients: an unexploited opportunity for ecology. *Ecology*, **71**, 1231–1237.

McGranahan, G. and Satterthwaite, D. (2003) Urban centers: an assessment of sustainability. *Annual Review of Environment and Resources*, **28**, 243–274.

McIntyre, N.E., Knowles-Yanez, K., and Hope, D. (2000) Urban ecology as an interdisciplinary field: Differences in the use of "urban" between the social and natural sciences. *Urban Ecosystems*, **4**, 5–24.

McKinney, M.L. (2002) Urbanization, biodiversity, and conservation. *BioScience*, **52**, 883–890.

McKinney, M.L. (2008) Effects of urbanization on species richness: A review of plants and animals. *Urban Ecosystems*, **11**, 161–176.

Murdoch, W.W. (1966) Community structure, population control and competition – a critique. *American Naturalist*, **100**, 219–226.

Padhi, R. (2007) Forced evictions and factory closures: rethinking citizenship and rights of working class women in Delhi. *Indian Journal of Gender Studies*, **14**, 73–92.

Park, R.E. and Burgess, E.W. (1967) *The City*, University of Chicago Press, Chicago.

Parris, K.M. (2006) Urban amphibian assemblages as metacommunities. *Journal of Animal Ecology*, **75**, 757–764.

Parris, K.M. and Hazell, D.L. (2005) Biotic effects of climate change in urban environments: the case of the grey-headed flying-fox (*Pteropus poliocephalus*) in Melbourne, Australia. *Biological Conservation*, **124**, 267–276.

Pickett, S.T.A. and P.S. White (eds). 1985. *The Ecology of Natural Disturbance as Patch Dynamics*. Academic Press, New York.

Pickett, S.T.A., Cadenasso, M.L., Grove, J.M. *et al.* (2001) Urban ecological systems: Linking terrestrial ecological, physical, and socioeconomic components of metropolitan areas. *Annual Review of Ecology and Systematics*, **32**, 127–157.

Riley, S.P. (2006) Spatial ecology of bobcats and gray foxes in urban and rural zones of a national park. *Journal of Wildlife Management*, **70**, 1425–1435.

Sewell, S.R. and Catterall, C.P. (1998) Bushland modification and styles of urban development: their effects on birds in south-east Queensland. *Wildlife Research*, **25**, 41–63.

Shochat, E., Lerman, S., Katti, M. *et al.* (2004) Linking optimal foraging behavior to bird community structure in an urban-desert landscape: Field experiments with artificial food patches. *American Naturalist*, **164**, 232–243.

Shochat, E., Warren, P.S., Faeth, S.H. *et al.* (2006) From patterns to emerging processes in mechanistic urban ecology. *Trends in Ecology and Evolution*, **21**, 186–191.

Smith, K.R., Brown, B.B., Yamada, I. *et al.* (2008) Walkability and body mass index. Density, design, and new diversity measures. *American Journal of Preventive Medicine*, **35**, 237–244.

Soule, D.C. (eds). 2006. *Urban Sprawl: A Comprehensive Reference Guide*. Greenwood Publishing Group, Westport.

Tait, C.J., Daniels, C.B., and Hill, R.S. (2005) Changes in species assemblages within the Adelaide metropolitan area, Australia, 1836-2002. *Ecological Applications*, **15**, 346–359.

Tansley, A.G. (1917) On competition between *Galium saxatile* L. (*G. hercynium* Weig.) and *Galium sylvestre* Poll. (*G. asperum* Schreb.) on different types of soil. *Journal of Ecology*, **5**, 173–179.

Thompson, K. and McCarthy, M.A. (2008) Traits of British alien and native urban plants. *Journal of Ecology*, **96**, 853–859.

Tibaijuka, A.K. (2005) *Report of the Fact-Finding Mission to Zimbabwe to assess the Scope and Impact of Operation Murambatsvina by the UN Special Envoy on Human Settlements Issues in Zimbabwe*, United Nations, New York.

UNFPA (2007) *State of World Population 2007. Unleashing the Potential of Urban Growth*, United Nations Population Fund, New York.

United Nations, Department of Economic and Social Affairs, Population Division (2014) *World Urbanization Prospects: The 2014 Revision*, United Nations, Geneva.

UN-Habitat (2006) *State of the World's Cities 2006/7: The Millennium Development Goals and Urban Sustainability*, Earthscan Publications, London.

Vähä-Piikkiö, I., Kurtto, A., and Hahkala, V. (2004) Species number, historical elements and protection of threatened species in the flora of Helsinki, Finland. *Landscape and Urban Planning*, **68**, 357–370.

van der Ree, R. and McCarthy, M.A. (2005) Inferring persistence of indigenous mammals in response to urbanization. *Animal Conservation*, **8**, 309–319.

Volterra, V. 1926. Variations and fluctuations of the numbers of individuals in animal species living together. Reprinted in R. N. Chapman. 1931. *Animal Ecology*. McGraw-Hill, New York.

Wackernagel, M. and Rees, W. (1996) *Our Ecological Footprint: Reducing Human Impact on the Earth*, New Society Publishers, Gabriola Island, Canada.

Wells, K.D. (1977) The social behavior of anuran amphibians. *Animal Behaviour*, **25**, 666–693.

Winterhalder, B. and Smith, E.A. (2000) Analyzing adaptive strategies: Human behavioral ecology at twenty-five. *Evolutionary Anthropology*, **9**, 51–72.

Williams, N.S.G., McDonnell, M.J., and Seager, E.J. (2005) Factors influencing the loss of indigenous grasslands in an urbanising landscape: a case study from Melbourne, Australia. *Landscape and Urban Planning*, **71**, 35–49.

Wittig, R. (2009) What is the main object of urban ecology? Determining demarcation using the example of research into urban flora, in *Ecology of Cities and Towns: A Comparative Approach* (eds M.J. McDonnell, A.K. Hahs, and J. Breuste), Cambridge University Press, Cambridge, pp. 523–529.

Zahavi, A. (1975) Mate selection: a selection for a handicap. *Journal of Theoretical Biology*, **53**, 205–214.

CHAPTER 2
Urban environments

2.1 Introduction

Human settlements arose around agricultural systems. The change from a nomadic, hunter-gatherer lifestyle to a more settled existence began in the Middle East during the Neolithic Period, around 9500 BCE (Bellwood 2004). By 3500 BCE, the first cities were established in Mesopotamia, in modern-day Iraq. Today, the size of human settlements varies from a cluster of dwellings in a village to towns, cities and megacities supporting more than 10 million people (Pearce 1996). Globally, there were 740 urban areas with a human population >500,000 in 2008, including 22 with a population >10 million (Cox 2008). The proportion of Earth's terrestrial surface that is classified as urban is currently estimated to be ~3% and rising (McGranahan et al. 2005; Seto et al. 2012). However, the ecological impact of cities and urban dwellers extends far beyond the urban boundary (Collins et al. 2000; Grimm et al. 2008a,b).

In Chapter 1, I briefly discussed the types of places where humans historically have chosen to settle. While these preferences account for a strong trend towards settlements in milder, coastal areas with high biodiversity (Luck et al. 2004), modern engineering and technology now enable people to live *en masse* in previously inhospitable landscapes such as deserts. For example, the Phoenix metropolitan area in Arizona, USA supports >4.2 million people while Dubai in the United Arab Emirates has a population >2.4 million (see Box 2.1). Coastal areas have the greatest share of urban land cover and the highest human-population density of the six ecosystems considered by the Millennium Ecosystem Assessment (McGranahan et al. 2007), followed by the cultivated zone and the inland water zone (habitats near rivers and lakes). Morris and Kingston (2002) proposed a theory that humans select their habitat to maximize individual fitness. Historically, this would have entailed selecting habitats with access to ample food and fresh water, secure shelter, and without significant threats from other humans or dangerous wild animals. The theory assumes that people have the option of moving from one place (or habitat) to another when they perceive a likely benefit of doing so.

Ecology of Urban Environments, First Edition. Kirsten M. Parris.

Box 2.1 Dubai and Phoenix – A Tale of Two Desert Cities

Dubai

Dubai is located in the United Arab Emirates (Figure 2.1). Its population grew from approximately 183,000 in 1975 to 2.4 million in 2015, a 13-fold increase in just 40 years (Dubai Statistics Center 2007, 2015). This population growth has been driven by immigration, particularly of expatriate workers, as well as natural increase (Pacione 2005). Dubai is currently experiencing a construction boom which includes grand projects such as the world's tallest building, the world's largest shopping mall, the Dubai Waterfront project (expected to house 1.2 million people) and a series of constructed off-shore islands shaped like palm trees and a map of the world (Pacione 2005; UAE Interact 2007). Tourism is also booming, with 4.36 million visitors to Dubai in 2012 (Dubai Statistics Centre 2012). In 2004, Dubai covered a geographic area of 605 km^2, with a further increase of up to 500 km^2 planned by 2015 (Pacione 2005). The population of the Dubai urban area is projected to grow to 3.3 million by 2025 (Cox 2008).

Figure 2.1 Skyscrapers in Dubai, UAE. Photograph by Jacqueline Schmid. https://pixabay.com/en/dubai-skyscraper-skyscrapers-639302/. Used under CC0 1.0 https://creativecommons.org/publicdomain/zero/1.0/deed.en.

Dubai is situated in the Arabian Desert. It has a hot, dry climate with an average annual rainfall of 94 mm and an average daily maximum temperature that exceeds 35 °C for 6 months of the year (WMO 2015a). Dubai has no rivers, so water for human consumption is generated from sea water at desalination plants. Electricity and water are largely produced in tandem; waste heat from the burning of natural gas to produce electricity is captured and used to produce steam for the desalination process (Dubai Electricity and Water Authority 2014). In 2004, total consumption of desalinated water in Dubai was 243.2 Gl, increasing to 482.7 Gl in 2014 (Dubai Electricity and Water Authority 2015a). Dubai's rapid development is also leading to increased demand for electricity. Electricity consumption more than doubled between 2004 and 2014, from 16.4 to 39.6 GWh (Dubai Electricity and Water Authority 2015b).

Phoenix

Phoenix is located in Arizona, USA (Figure 2.2). The population of metropolitan Phoenix grew from 330,000 people in 1950 to 4.2 million in 2012 (Brazel et al. 2007; Guhathakurta and Gober 2007; Shrestha et al. 2012). More than 60,000 permits were issued for the construction of new homes in 2004 alone, both on the city's fringe and within the boundaries of existing development (Brazel et al. 2007). The population of Maricopa County (which is largely determined by the population of metropolitan Phoenix) is predicted to reach 6.2 million by 2030 and 7.7 million by 2050 (Arizona Department of Economic Security 2006). This rapid urban growth has been driven by an increased availability of water, an expansion of industry, and an influx of people seeking the lifestyle of a desert oasis in the "Valley of the Sun".

Figure 2.2 Midtown skyline of Phoenix, Arizona. Picture has been cropped and converted to black and white. Photograph by Sean Horan. Used under CC-BY-2.0 https://creativecommons.org/licenses/by/2.0/.

Phoenix is situated in the Sonoran desert. It has a hot, dry climate, with an average daily maximum temperature that exceeds 35 °C for 4 months of the year, and an average annual rainfall of 208 mm (WMO 2015b). Energy for the city is provided by power stations that burn coal and natural gas, a nuclear power station, hydroelectric dams and some solar power stations (US EIA 2015). Water for human consumption, gardens, swimming pools, golf courses and artificial lakes is diverted from upstream watersheds (catchments) of the Salt, Verde and Colorado Rivers or from underground aquifers (Balling et al. 2008). Between 1995 and 2003, the average household in Phoenix used 1788 litres of water per day, much of it for outdoor purposes, such as watering gardens and filling swimming pools (Balling et al. 2008). However, per-capita water use in Phoenix has been falling in recent years following a variety of water-conservation initiatives (Gammage et al. 2011).

Desert living and the ecological footprint

The World Wide Fund for Nature's Living Planet Report (WWF 2014) identified Kuwait, Qatar and the United Arab Emirates as the three countries with the largest per-capita ecological footprint in the world for 2010. Kuwait had a footprint of 10.4 global hectares (gha) per person, compared to the United States of America with a footprint ~7 gha/person (ranked 8th in the world) and Jamaica with a footprint ~2 gha/person (ranked 76th in the world). A country's footprint is the sum of all the land and fishing grounds required to produce the resources it consumes (e.g., fuel, food, fibre and timber), to absorb the wastes it creates, and to provide space for its infrastructure (but see criticism of the ecological footprint concept; Blomqvist et al. 2013). In 2010, the Earth's average biocapacity or total productive area was 1.7 gha/person, where a global hectare is a hectare with world-average ability to produce resources and absorb wastes (WWF 2014). The global ecological footprint was 2.6 gha/person in 2010, indicating that humans are living beyond our environmental means – we would need >1.5 Earths to sustain our current demand for ecological resources. Large cities in the desert, such as Dubai and Phoenix, make a disproportionately large per-capita contribution to the ecological footprint of their respective countries, because of the energy used to cool homes, cars, office buildings and shopping malls, and the water used to maintain the oasis culture of gardens, golf courses, lakes and swimming pools. Construction of cities increases the local temperature above that of the surrounding area, a phenomenon known as the urban heat-island effect (see Box 2.2). As cities get larger and denser, they also get warmer. Therefore, in a big city in a hot desert climate, even more energy is required for cooling buildings and cars, and even more water is used to maintain gardens, parks and swimming pools (Baker et al. 2002; Guhathakurta and Gober 2007; Grimm et al. 2008a).

Where urban living provides access to adequate housing, clean drinking water and health care, the fitness of humans is higher in cities than in rural areas (Van de Poel et al. 2009; WHO and UN-Habitat 2010; World Bank 2013). However, the health of people living in urban slums can sometimes be worse than that of their rural counterparts (and is markedly worse than that of people living in well-planned parts of cities), with higher rates of certain diseases, malnutrition and infant mortality (Dyson 2003; UNFPA 2007; WHO and UN-Habitat 2010; World Bank 2013). Many people undoubtedly move from rural to urban habitats in the hope of improving their personal circumstances; whether they have the option to move back again if this hope is not realized is questionable.

The construction of cities, the pace of urban growth and the form of urban development (e.g., urban sprawl, high-density housing, informal settlements) are shaped by a variety of economic, social and cultural factors which interact with each other and change over time (Lyon and Driskell 2011; Macionis and Parrillo 2012). This book is less concerned with the socio-economic drivers of urbanization (e.g., Henry et al., 2003; Alig et al. 2004; Liu et al. 2005) than with its consequences

Box 2.2 The Urban Heat-Island Effect

The urban heat-island effect was first documented in the early 19th century by the amateur meteorologist Luke Howard, following his observations that overnight minimum temperatures in central London were higher than those in the surrounding countryside (Howard 1833). Heat islands form in urban areas because the materials used to construct buildings, roads and pavements absorb and retain more heat from the sun than natural vegetated surfaces. The heat that is absorbed during the day is then released back into the atmosphere at night. Heat produced by industry, vehicles and the heating or cooling (air conditioning) of buildings contributes to the urban heat island, as do particulate air pollution and reduced wind speeds in cities (Gartland 2008). Replacement of vegetation with impermeable surfaces also reduces the amount of thermal energy lost from cities through evaporation, further increasing the storage of heat during the day and its release at night (Gartland 2008). Brazel et al. (2007) identified a 1.4 °C increase in the June mean minimum temperature with every 1000 new homes constructed in a 1-km radius of a weather station between 1990 and 2004 in Phoenix, Arizona. In Beijing, China, the percentage of impervious surface explains around 80% of local spatial variation in land surface temperature during summer (Ouyang et al. 2008).

In temperate climates, the heat-island effect is most pronounced on clear, still nights in winter. For example, a transect of the temperature through Melbourne, Australia on a clear winter's evening in August 1992 revealed a peak warming effect of 7.1 °C in the central business district compared to the rural area on the western boundary of the city (Torok et al. 2001). Similar patterns have been observed in tropical Singapore, where the urban heat-island effect is most pronounced on clear, calm evenings in the drier months of the year (Chow and Roth 2006). The urban heat-island effect can substantially increase late afternoon and overnight temperatures in summer, contributing to human discomfort, illness and death in hot climates (Brazel et al. 2000; Harlan et al. 2007; Tan et al. 2010). Urban heat islands can be mitigated in a variety of ways, such as through the use of cool roofing and cool paving materials, and by increasing vegetation cover via landscaping and the establishment of green roofs and walls (Gartland 2008; Gago et al. 2013). Cooling cities in this way can provide a range of benefits to the people who live there, including energy savings, cleaner air, and improved health and wellbeing (Gartland 2008). Increasing the cover of vegetation in cities may also benefit local biodiversity.

for populations, species, ecological communities, ecosystem function, and the health and wellbeing of human city-dwellers. In this chapter, I examine the processes of urbanization, the characteristics of urban habitats, and the ways in which they differ from rural or non-urban habitats.

2.2 Primary biophysical processes associated with urbanization

Despite differences in geology, vegetation and climate in different parts of the globe, the biophysical processes associated with the construction of towns and cities are often similar. Urban habitats also share many physical features,

Table 2.1 The primary biophysical processes of urbanization and the secondary biophysical processes that arise from these.

Primary Processes	Secondary Processes
1. Removal of existing vegetation	Habitat loss, fragmentation and isolation (arising from 1, 2, 4, 5)
2. Construction of buildings, roads, lights, drains and other urban infrastructure	Climatic changes (arising from 1, 2, 3, 4)
3. Replacement of permeable with impermeable surfaces	Altered noise regimes (arising from 1, 2, 3)
4. Reduction in the area of open space	Altered light regimes (arising from 2, 3)
5. Modification or destruction of aquatic habitats	Altered hydrological regimes (arising from 1, 2, 3, 4, 5)
6. Production of pollution and waste	Pollution of air, water and soil (arising from 1, 2, 3, 6)

presenting similar opportunities for and challenges to the persistence of non-human species. The primary biophysical processes of urbanization include: removal of existing vegetation; construction of buildings, roads, lights, drains, fences and other urban infrastructure; replacement of permeable with impermeable surfaces; reduction in the area of open space; modification or destruction of aquatic habitats, such as ponds, streams and rivers; and production of pollution and waste (Table 2.1). I discuss each of these in detail below.

2.2.1 Removal of existing vegetation

When a town or city is constructed, much of the existing vegetation is removed. Depending on the climate, soil type and land-use history of an urbanizing area, this existing vegetation can be indigenous primary forest, woodland, heathland, grassland or sparse desert vegetation; secondary or re-growth vegetation; or weeds and pasture grasses. In general, the cover of vegetation in a town or city declines as human population density and/or the density of development increases (e.g., Pauchard et al. 2006; Kromroy et al. 2007). Note that urban density has been defined and measured in many ways, depending on the entity or question of interest (e.g., the number of people, dwellings or jobs per unit area; total floor area or total open space per unit area; Dovey and Pafka 2014). In a study of 12 cities built in naturally forested areas in the USA, average tree cover was highest in parks and vacant land, intermediate in residential areas, and lowest in commercial and industrial areas (Nowak et al. 1996; Figure 2.3). Er et al. (2005) estimated that 87% of the forested area in Vancouver, Canada was replaced by urban development between 1859 and 1999. Interestingly, urban expansion can increase the net cover of native vegetation in non-urban areas. For example, forest cover on the island of Puerto Rico increased from 28 to 40% between 1991 and 2000, with a shift from rural to urban living and a concomitant abandonment of agricultural

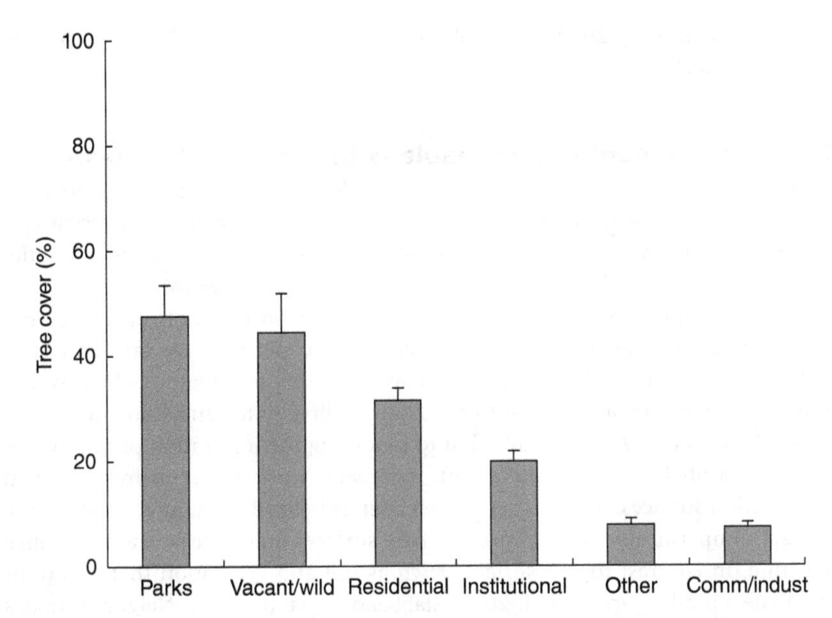

Figure 2.3 Tree cover (%) in different urban land-use types, averaged across 12 cities built in naturally forested areas in the USA. Vacant/wild = vacant lots or wild lands, institutional = institutional areas such as hospitals and schools, other = agricultural land, orchards, roads and airports; comm/indust = commercial and industrial land. The error bars represent standard errors. Data from Nowak et al. (1996).

fields (Parés-Ramos et al. 2008). Planting of trees in gardens and streets may lead to an increase in tree cover in cities with time since development, partially compensating for the initial loss of vegetation with urbanization. However, the species planted are often not indigenous to the local area and do not replace the original vegetation community (Tait et al. 2005; Shukuroglou and McCarthy 2006).

2.2.2 Construction of buildings, roads and other urban infrastructure

Perhaps the most obvious physical process associated with urbanization is the construction of dwellings and other buildings, often with supporting infrastructure such as fences, paved roads and footpaths, electricity and water supply, storm-water drains, sewers and street lights. Formal settlements tend to have well-constructed houses and well-developed infrastructure. Conversely, informal settlements are often comprised of makeshift dwellings without electricity, sanitation, a supply of clean water or well-defined roads. The central business district of many cities (also known as the inner city or urban core) is comprised of a dense array of office blocks, apartment buildings, shops, hotels, law courts and places of worship. Residential (suburban) areas tend to support a lower density of buildings (McKinney 2002), particularly in affluent neighbourhoods with large residential blocks.

2.2.3 Replacement of permeable with impermeable surfaces

Permeable surfaces are those that allow the infiltration of water, while impermeable (or impervious) surfaces are those that do not. When cities are constructed, the area covered with impermeable surfaces increases substantially as soil, sand, grass and trees are replaced with buildings and paved surfaces (McKinney 2002). When precipitation falls on a city, much of it lands on the roofs of houses and other structures, asphalt roads and car parks, or concrete driveways and footpaths. In many cases, storm-water systems carry this rain water (which would otherwise have been absorbed into the ground) directly to canals, streams or the ocean (Walsh et al. 2004). In addition to hindering the infiltration of rain water, buildings, asphalt and concrete absorb more shortwave radiation from the sun than natural surfaces, and this is released later as heat (Bridgman et al. 1995; see Box 2.2). Impermeable surfaces are also hard surfaces that increase the reflectance of sound, in contrast to soft surfaces such as soil and vegetation that absorb or attenuate sound (Warren et al. 2006; Slabbekoorn et al. 2007). Neighbourhoods with a high proportion of impermeable surfaces therefore tend to be warmer and noisier than those with a lower proportion. Finally, paved surfaces separate the soil from air and water, interfering with biotic processes, such as soil nutrient cycling and gas exchange (Scalenghe and Marsan 2008), and preventing penetration of the ground by plant roots or digging and burrowing animals.

Scalenghe and Marsan (2008) estimated that 9% of the terrestrial surface of Europe is covered with an impermeable material, while the global area of impermeable surface probably exceeds 500,000 km^2 or 0.34% (Elvidge et al. 2007). Within a city, the proportion of impermeable surface generally increases with urban density. For example, a recent study found the average cover of impermeable surface in six concentric zones in Beijing, China varied from 67.3% in the centre of the city to 9.3% in the outermost zone (Xiao et al. 2007). Across the whole study area of 4084 km^2, the average cover of impermeable surface was 20.8%, but significant sections of the inner-urban zones had an impermeable surface cover >90% (Xiao et al. 2007). In a survey of 37 cities in the USA, impermeable surface cover was highest in commercial and industrial neighbourhoods, lower in residential neighbourhoods and lowest in parks and vacant lands (Nowak et al. 1996).

2.2.4 Reduction in the area of open space

Open spaces (also known as green spaces or green open spaces) have few buildings or roads, and include city parks, nature reserves, wetlands, river corridors, reservoirs, market gardens, airports, sports fields, golf courses, wastelands and

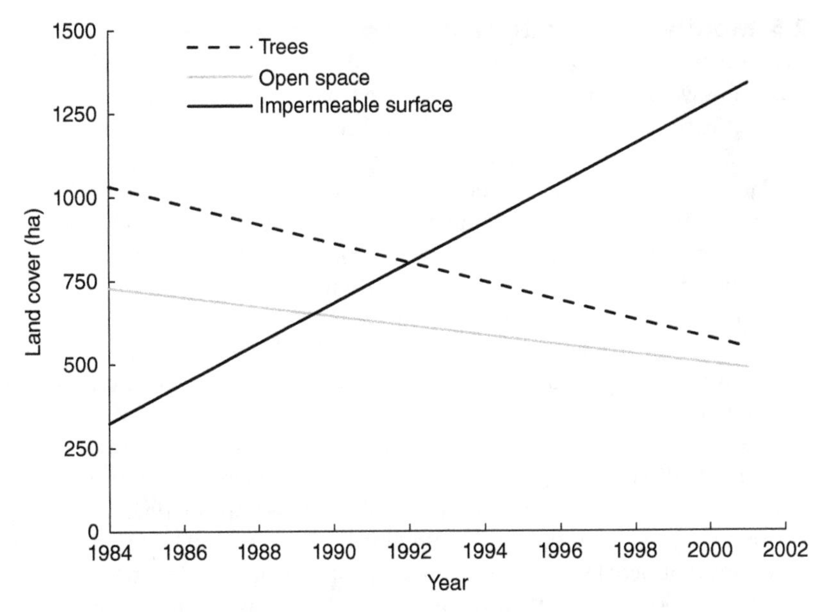

Figure 2.4 Land-cover change in Cornelius, North Carolina, USA resulting from urban expansion, 1984–2001. Impermeable-surface cover increases while the cover of trees and open space decreases. Data from American Forests (2003).

vacant lots (Leary and McDonnell 2001; Forman 2008, 2014; Ishan et al. 2014). Open space declines with the construction of buildings, roads and other urban infrastructure and is inversely proportional to the cover of impermeable surface (e.g., American Forests 2003; Figure 2.4). Public open spaces in cities provide opportunities for human recreation, as well as habitat for plants and animals. But as the human population of a city increases, areas of open space are often lost to development. For example, 14,000 km^2 of urban open space was lost in the contiguous United States between 1990 and 2000, and the area lost per city was strongly correlated with human population growth ($r^2 = 0.72$; McDonald et al. 2010). Species–area relationships mean that larger areas of open space often support a higher diversity of species, including animals with relatively complex life histories, large area requirements, and those that are sensitive to disturbance by humans or noise (Fernández-Juricic 2000; Angold et al. 2006; Murgui 2007). However, in a survey of 38 urban regions around the world, only 16 had a single area of open space larger than 4 km^2 (Forman 2008). In 11 regions, the largest area of open space was <1 km^2. Nevertheless, even small areas of open space can contribute to the conservation of native plants and animals in cities (Gibb and Hochuli 2002; Rosensweig 2003; Fuller et al. 2007) and provide important benefits for human health and wellbeing (Matsuoka and Kaplan 2008).

2.2.5 Modification or destruction of aquatic habitats

Many aquatic habitats are altered or lost during the construction of cities: ponds and swamps are filled or constrained tightly within engineered walls; streams and rivers are straightened, controlled with locks and weirs, diverted underground or paved with concrete; ephemeral wetlands and floodplains may be replaced with manicured parklands (Figure 2.5). Some of these measures are taken in an attempt to control flooding, or to remove the breeding grounds of pathogen-bearing mosquitoes (Meredith 2005). Wetlands are often demolished to make way for urban development (e.g., Kentula et al. 2004; Pauchard et al. 2006). For example, urbanization destroyed 1734 ha (23%) of wetlands in Concepción, Chile between 1975 and 2000 (Pauchard et al. 2006). Most of these wetlands were drained and filled to provide space for residential and industrial developments.

Where wetlands remain in urban environments, their structure can be altered by changed hydrological regimes and/or land management practices. Managers may mow the vegetation in and around unkempt swamps or riparian areas to make them appear neat and well-maintained (Nassauer 2004). In addition to the changes that people deliberately make to aquatic habitats in cities, unintentional changes occur with urbanization. These commonly include increased suspended sediment loads in streams and rivers, erosion of banks and enlargement of river channels (Suren et al. 2005; Chin 2006). In some cities, restoration projects attempt to return water courses and wetlands closer to their natural state, for the

Figure 2.5 The River Ouse in York, UK, showing the confined stream channel. Photograph by Kirsten M. Parris.

benefit of humans and non-humans alike. These projects include revegetation of stream corridors and wetlands, and experiments to replace concrete stream channels with natural banks (e.g., Hynes et al. 2004; Suren et al. 2005).

2.2.6 Production of pollution and waste

Human beings and industry produce pollution and waste, which are concentrated in areas where large numbers of people live, travel and work. Commonly, they include human waste, household and industrial rubbish, atmospheric pollution from industry and car exhaust, household and industrial chemicals, and nutrients such as nitrogen and phosphorus. Cities of the Minoan Civilisation on Crete (3000–1100 BCE) and the Harappa/Indus Valley Civilisation in present-day Pakistan, India and Afghanistan (2600–1900 BCE) had the first known urban sewerage and sanitation systems, including flushing toilets, drains from each house that connected to brick-lined sewers, rubbish chutes built into the walls of houses, and public rubbish bins (Angelakis et al. 2014; Khan 2014). Today, systems for managing pollution and waste vary greatly between cities, depending on levels of infrastructure, environmental-protection legislation and compliance, and the extent of informal settlements.

Efficient collection, transport and treatment of human waste are of prime importance for the health of urban dwellers, particularly where population densities are high (McMichael 2000). Cities in developed countries tend to have more comprehensive systems of sewers and more sophisticated sewage treatment plants than those in developing countries (Harada et al. 2008; Beyene et al. 2009), as well as stronger legislative controls on the production of atmospheric pollution and the disposal of household and industrial waste (e.g., Pargal et al. 1997; Hettige et al. 1998; UNEP 2005). Concentrations of atmospheric pollutants such as particulate matter vary dramatically between cities and countries (Figure 2.6). For example, in 2010, the mean annual concentration of coarse particulate matter (PM_{10}) in the atmosphere was 286 $\mu g\, m^{-3}$ in New Delhi, India and 42 $\mu g\, m^{-3}$ in La Paz, Bolivia, measured across 5 and 4 stations, respectively (WHO 2014a). The World Health Organization air quality guidelines for PM_{10} are an annual mean of 20 $\mu g\, m^{-3}$ (WHO 2014b).

In 2004, an estimated 1.3 Pg (1.3 billion metric tonnes) of municipal waste (household garbage) were collected worldwide, not including waste from rural parts of non-OECD countries such as India and China (Lacoste and Chalmin 2007). This is expected to increase to 2.2 Pg by 2025 (Hoornweg and Bhada-Tata 2012). The USA has the highest per capita rate of waste production in the world (Lacoste and Chalmin 2007), with an average of 725 kg of municipal solid waste produced per person in 2012 (US EPA 2014; Figure 2.7a). Interestingly, while per capita waste production in the USA has been constant or declining since 1990, the proportion of this waste that is recycled or composted has increased from 16.0 to 34.5%, reducing the volume of waste going to landfill (US EPA 2014; Figure 2.7b). In Nairobi, Kenya, each resident produces an average of 183 kg of solid municipal

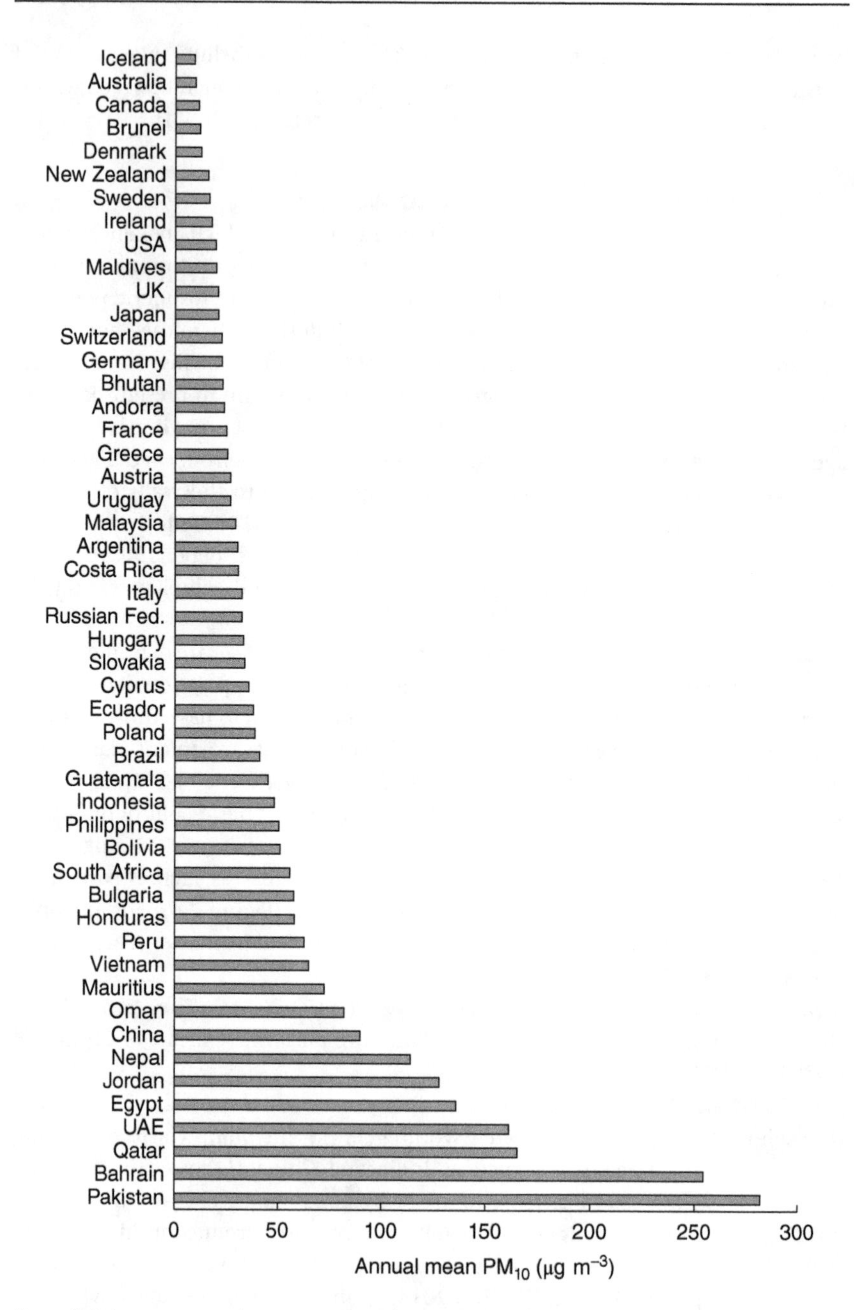

Figure 2.6 Average annual concentration of coarse particulate matter (PM_{10}) in the atmosphere (μg m^{-3}), recorded in 50 countries. Data from WHO 2014.

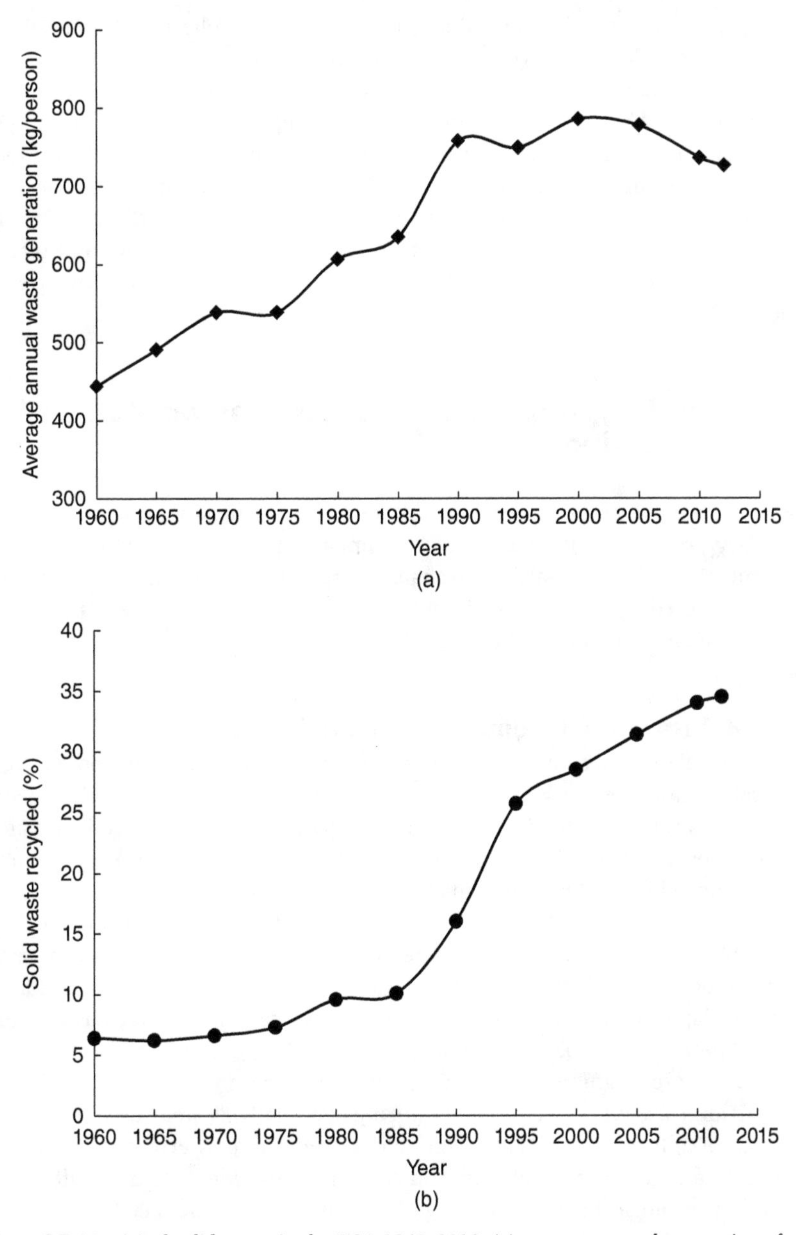

Figure 2.7 Municipal solid waste in the USA 1960–2012: (a) average annual generation of waste (kg/person/year); b) percentage of waste recycled. Data from US EPA 2014.

waste per year, but only about one quarter of this is collected (UNEP 2005). Formal waste collection services are almost non-existent in the city's slums and unplanned settlements. As a consequence, Nairobi has a substantial problem with littering and pollution from the uncontrolled dumping of waste (UNEP 2005). Non-biodegradable plastic bags are particularly problematic, as they can block gutters and storm-water drains, provide breeding habitat for malaria-bearing mosquitoes, and reduce the productivity of agricultural fields (Njeru 2006). In 2007, the Kenyan government banned the use of thin plastic bags and placed a high tax on thicker plastic bags in an effort to reduce their use (BBC News, June 14, 2007).

2.3 Secondary biophysical processes associated with urbanization

The primary biophysical processes that occur with urbanization lead to a range of secondary processes that have a profound impact on the ecology of urban environments. These include habitat loss, habitat fragmentation and isolation, climatic changes, altered hydrological and light regimes, increased noise levels, and the pollution of air, water and soil (Table 2.1).

2.3.1 Habitat loss, fragmentation and isolation

As outlined above, substantial areas of habitat for non-human species, both terrestrial and aquatic, are lost when cities are constructed. In many cases, the habitat that remains is fragmented into small areas or patches, which tend to be isolated from each other by an inhospitable landscape of urban infrastructure, people and fast-moving vehicles. Over time, small patches of remnant habitats in urban environments are altered further. Their high perimeter to area ratio means they are susceptible to edge effects, such as altered light, thermal and wind regimes, invasion by weeds, and trampling by humans (Cilliers et al. 2008; Hamberg et al. 2008). Urban wetlands are also readily invaded by introduced animals such as fish and turtles (Spinks et al. 2003; Riley et al. 2005; Bury 2008). In general, urban landscapes are more fragmented than rural landscapes, with many small patches of different land-use types or biotopes occurring in a relatively small area (e.g., Luck and Wu 2002; Breuste et al. 2008). Biotope is a term used by ecologists in Europe to describe an area of relatively uniform environmental condition associated with a particular ecological community (e.g., Löfvenhaft et al. 2002, 2004).

2.3.2 Climatic changes

Construction of a city can cause pronounced climatic changes. While the best known of these is the urban heat-island effect (see Box 2.2), urbanization can also increase cloudiness, reduce solar radiation, alter rainfall, humidity and wind patterns, and enhance thunderstorm activity (Bridgman et al. 1995; Changnon 2001; Sturman and Tapper 2006). Human activities such as manufacturing, power generation, evaporative cooling and watering of parks and gardens release moisture into the atmosphere, increasing the absolute humidity over cities (but note that relative humidity may be lower because the air is warmer and can therefore hold more moisture; Bridgman et al. 1995). In still conditions, the combination of increased humidity and particulate pollution in the atmosphere – which provides nuclei for condensation and the formation of cloud droplets – enhances the development of fog. During winter, cities can have up to twice as many foggy days as the surrounding rural area (Landsberg 1981). In conditions that favour vertical convection, cloud formation is enhanced by atmospheric moisture, air pollution, heat and reduced wind speeds in the city (Bridgman et al. 1995). This increases cloudiness and rainfall and reduces solar radiation, although particulate air pollution also reduces solar radiation in the absence of cloud (Sturman and Tapper 2006). In windy conditions, the diverse heights (or roughness) of city buildings can reduce wind speeds below those in the surrounding rural area, but tall buildings can also funnel wind and create local "wind-tunnel" effects (Bridgman et al. 1995).

2.3.3 Altered hydrological regimes

Replacement of permeable with impermeable surfaces in urban watersheds (catchments) and construction of efficient storm-water drainage systems dramatically change the hydrology of cities. Rain that falls upon impermeable roofs and paved surfaces is quickly transported to streams via storm-water systems, leading to a sudden increase in the rate of stream flow – even for relatively small rainfall events – followed by a rapid decline (Walsh et al. 2005; Figure 2.8). Conversely, reduced infiltration of rain water lowers soil moisture and the flow of ground water, which may reduce the base flow of a stream below that of a comparable stream in a rural watershed. Thus, urban streams are characterized by lower base flows and higher peak flows than rural streams, as well as flash flood events (Figure 2.8). These short-lived, high-flow events tend to alter the geomorphology of streams by eroding banks, incising channels, and reducing structural complexity (Walsh et al. 2005). Changes to flow regimes and the physical habitat of urban streams have important consequences for stream biota, including invertebrates, fish, amphibians and reptiles.

2.3.4 Pollution of air, water and soil

Air pollution from industrial emissions, car exhaust and solid-fuel fires is often the most obvious form of pollution in urban environments, but polluted soils, sediments and waterways are also very common. Depending on the type and concentration of pollutants (e.g., diesel particulates, carbon monoxide, ozone, benzene, heavy metals, phosphorus, nitrogen, pesticide and herbicide residues, municipal waste and/or human waste), urban pollution can adversely affect the health of humans as well as other animals and plants. For example, exposure to air pollution

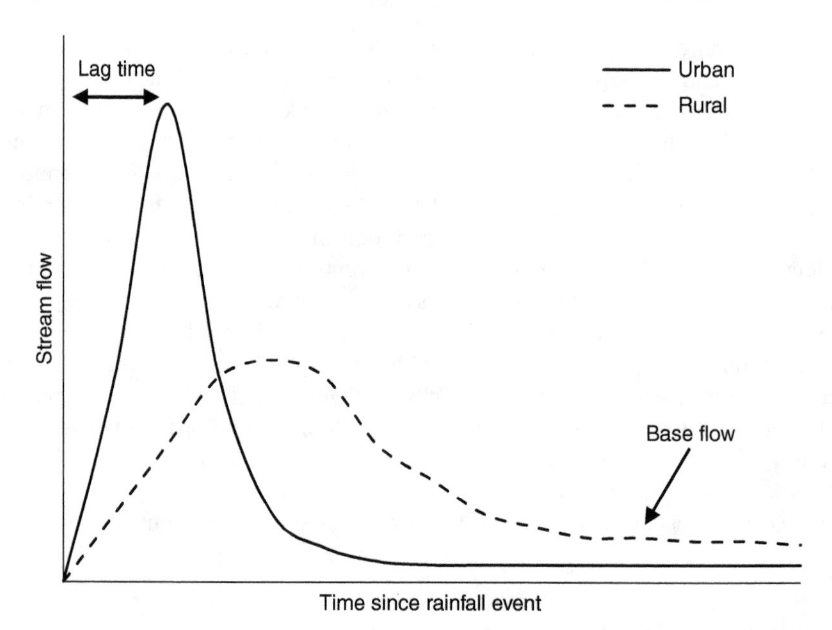

Figure 2.8 Schematic diagram of an urban and a rural hydrograph, showing stream flow versus time since a rainfall event. The urban curve has a shorter lag time, a higher peak, a faster subsidence and a lower base flow than the rural curve, reflecting the characteristic "flashiness" of urban streams.

increases the risk of cardiovascular disease, respiratory diseases such as asthma and emphysema, cancer, and infertility in humans (Brook 2008; Krivoshto et al. 2008). The herbicide glyphosate, which is commonly used in both urban and rural areas, can kill amphibians at the larval and adult stage (Relyea 2005). Deposition of atmospheric nitrogen downwind of urban centres alters the structure of plant and microbial communities by favouring some species and disadvantaging others (Fenn et al. 2003). In the western United States, high levels of soil nitrogen have enhanced invasion of plant communities by exotic grasses, which then outcompete native herbs and forbs including the host plants of threatened butterfly species (Fenn et al. 2003). Municipal waste (household garbage) provides food for a wide variety of animal species, both native and introduced, in cities around the world (e.g., Beckmann and Berger 2003).

2.3.5 Altered noise and light regimes

In general, cities are noisier than rural areas, but urbanization also changes noise regimes in other ways. The frequency (spectral) distribution of urban noise differs from that of rural areas, as does the timing and persistence of noise throughout the day, and the reverberation and attenuation of noise across the structurally complex

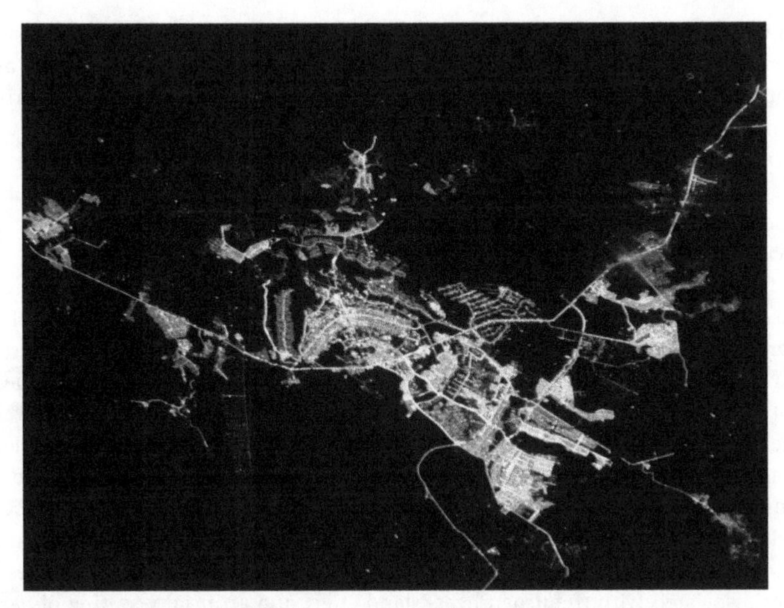

Figure 2.9 Ecological light pollution in Brasilia, Brazil as seen at night from space. Photograph courtesy of the Earth Science and Remote Sensing Unit, NASA Johnson Space Center.

urban landscape (Patricelli and Blickley 2006; Warren et al. 2006). Much terrestrial urban noise comes from vehicular traffic on roads, but other sources such as industry, construction, passing aeroplanes and trains, lawn mowers, leaf blowers and amplified music may all contribute (Warren et al. 2006). In aquatic habitats such as rivers, lakes, estuaries and harbours, underwater noise from boat traffic and the construction of bridges and causeways can be substantial (Lesage et al. 1999; Haviland-Howell et al. 2007; Graham and Cooke 2008; Bailey et al. 2010). Cities are also lighter than rural areas. Ecological light pollution occurs when artificial lights alter the natural diurnal cycle of light and dark in ecosystems (Longcore and Rich 2004; Gaston et al. 2013). Ecological light pollution in the cities of the world is so pronounced that it can be seen clearly from space (Figure 2.9).

2.4 Stochasticity in urban environments

In any habitat, environmental conditions such as temperature, precipitation, wind, soil moisture levels and the availability of food and other resources may vary from day to day, season to season and year to year. Environmental stochasticity can be defined as small-to moderate-scale, random variations in environmental conditions over time that affect the birth and death rates of individuals in a population (May 1974; Lande 1993). It is generally distinguished from demographic stochasticity, which arises from chance births and deaths within a population, and catastrophes, which are large-scale environmental perturbations that lead to dramatic population declines (Lande 1993). I discuss the impact of different forms of stochasticity on populations further in Chapter 3, but it is worth considering here whether urban environments might be expected to experience more or less environmental stochasticity than non-urban environments.

The various biophysical processes associated with urbanization do not proceed in a deterministic way and, as mentioned in Section 2.3.1, urban habitats are often characterized by spatial variability or patchiness (Niemela 1999; Fernandez-Juricic and Jokimaki 2001; Luck and Wu 2002). But this does not necessarily mean that environmental conditions in a given location within a city are highly variable over time. The frequent disturbance of urban habitats by humans may lead to increased temporal variation in the environmental conditions experienced by particular groups of species, such as grasses, herbs or ground-dwelling arthropods (Rebele 1994; Sattler et al. 2010). However, some environmental conditions may actually be less variable in cities than in nearby rural areas. Climatic changes in cities associated with the urban heat-island effect and artificial watering of parks and gardens can reduce diurnal and seasonal variations in temperature, humidity and soil moisture (Oke 1982; Parris & Hazell 2005). Furthermore, seasonal variation in the availability of food for generalist species such as granivorous birds, frugivorous bats and scavenging mammals may be substantially lower in urban than rural areas, leading to a year-round supply of suitable food and thus higher population densities (Beckmann and Berger 2003; Contesse et al. 2004; Shochat 2004; Williams et al. 2006). Increased or reduced levels of environmental stochasticity experienced by different species following urbanization may contribute to the persistence or local extinction of their populations over time.

2.5 Summary

It is clear that urbanization can destroy existing habitats, alter others, and create novel habitats without obvious parallel in more rural or pristine environments. However, the primary and secondary biophysical processes associated with the construction of towns and cities are often similar throughout the world, regardless of local vegetation, climate and landforms. How do microbes, fungi, plants and animals respond to these processes? How do these processes affect the ecology of populations, species and communities? To what extent are ecosystem processes maintained in cities? How can we conserve native biodiversity in urban environments? How does city life impact on the health and wellbeing of humans? Where do the needs of humans and other species conflict or overlap? And how can we plan and construct the cities of the future to improve habitat for humans and non-humans alike? I address these questions in the following chapters.

Study questions

1 What is an ecological footprint? What influences the ecological footprint of a city?
2 Discuss the primary biophysical processes of urbanization. Which processes do you think have the greatest environmental impact?
3 What are the key secondary biophysical processes of urbanization? Consider how these might affect humans and other species in cities.
4 Explain the urban heat-island effect and its impact on minimum and maximum air temperatures in cities.
5 What is environmental stochasticity? Is it likely to be high or low in urban environments?

References

Alig, R.J., Kline, J.D., and Lichtenstein, M. (2004) Urbanization on the US landscape: looking ahead in the 21st century. *Landscape and Urban Planning*, **69**, 219–234.

American Forests (2003) *Urban Ecosystem Analysis: Mecklenburg County, North Carolina*. American Forests, Washington DC. http://charmeck.org/city/charlotte/epm/services/landdevelopment/trees/treecommission/documents/charlotte_final.pdf (accessed on 8/07/2015).

Angelakis, A.N., Kavoulaki, E., and Dialynas, E.G. (2014) Sanitation and wastewater technologies in Minoan Era, in *Evolution of Sanitation and Wastewater Technologies Through the Centuries* (eds A.N. Angelakis and J.B. Rose), IWA Publishing, London, pp. 1–24.

Angold, P.G., Sadler, J.P., Hill, M.O. *et al.* (2006) Biodiversity in urban habitat patches. *Science of the Total Environment*, **360**, 196–204.

Arizona Department of Economic Security (2006) *Arizona County Population Projections*, State Data Centre, Phoenix, Arizona, USA.

Bailey, H., Senior, B., Simmons, D. *et al.* (2010) Assessing underwater noise levels during pile-driving at an offshore windfarm and its potential effects on marine mammals. *Marine Pollution Bulletin*, **60**, 888–897.

Baker, L.A., Brazel, A.J., Selover, N. *et al.* (2002) Urbanization and warming of Phoenix (Arizona, USA): Impacts, feedbacks and mitigation. *Urban Ecosystems*, **6**, 183–203.

Balling, R.C., Gober, P., and Jones, N. (2008) Sensitivity of residential water consumption to variations in climate: An intraurban analysis of Phoenix, Arizona. *Water Resources Research*, **44**, 1–11.

BBC News, June 14, 2007; http://news.bbc.co.uk/2/hi/africa/6754127.stm (accessed on 02/11/2015).

Beckmann, J. and Berger, J. (2003) Rapid ecological and behavioural changes in carnivores: the responses of black bears (*Ursus americanus*) to altered food. *Journal of Zoology*, **261**, 207–212.

Bellwood, P. (2004) *The First Farmers: The Origins of Agricultural Societies*, Blackwell Publishing, Malden.

Beyene, A., Legesse, W., Triest, L. *et al.* (2009) Urban impact on ecological integrity of nearby rivers in developing countries: the Borkena River in highland Ethiopia. *Environmental Monitoring and Assessment*, **153**, 461–476.

Blomqvist, L., Brook, B.W., Ellis, E.C. *et al.* (2013) Does the shoe fit? Real versus imagined ecological footprints. *PLoS Biology* **11** (11) e1001700 doi:10.1371/journal.pbio.1001700.

Brazel, A., Gober, P., Lee, S. *et al.* (2007) Determinants of changes in the regional urban heat island in metropolitan Phoenix (Arizona, USA) between 1990 and 2004. *Climate Research*, **33**, 171–182.

Brazel, A., Selover, N., Vose, R. *et al.* (2000) The tale of two climates–Baltimore and Phoenix urban LTER sites. *Climate Research*, **15**, 123–135.

Breuste, J.H., Niemelä, J., and Snep, R.P.H. (2008) Applying landscape ecological principles in urban environments. *Landscape Ecology*, **23**, 1139–1142.

Bridgman, H., Warner, R., and Dodson, J. (1995) *Urban Biophysical Environments*, Oxford University Press, Melbourne.

Brook, R.D. (2008) Cardiovascular effects of air pollution. *Clinical Science*, **115**, 175–187.

Bury, B. (2008) Do urban areas favour introduced turtles in western North America?, in *Urban Herpetology* (eds J.C. Mitchell, R.E.J. Brown, and B. Bartholomew), Society for the Study of Amphibians and Reptiles, Salt Lake City, pp. 343–345.

Changnon, S.A. (2001) Assessment of historical thunderstorm data for urban effects: The Chicago case. *Climatic Change*, **49**, 161–169.

Chin, A. (2006) Urban transformation of river landscapes in a global context. *Geomorphology*, **79**, 460–487.

Chow, W.T.L. and Roth, M. (2006) Temporal dynamics of the urban heat island of Singapore. *International Journal of Climatology*, **26**, 2243–2260.

Cilliers, S.S., Williams, N.S.G., and Barnard, F.J. (2008) Patterns of exotic plant invasions in fragmented urban and rural grasslands across continents. *Landscape Ecology*, **23**, 1243–1256.

Collins, J.P., Kinzig, A., Grimm, N.B. *et al.* (2000) A new urban ecology: Modeling human communities as integral parts of ecosystems poses special problems for the development and testing of ecological theory. *American Scientist*, **88**, 416–425.

Contesse, P., Hegglin, D., Gloor, S. *et al.* (2004) The diet of urban foxes (*Vulpes vulpes*) and the availability of anthropogenic food in the city of Zurich, Switzerland. *Mammalian Biology*, **69**, 81–95.

Cox, W. (2008) *Demographia World Urban Areas (World Agglomerations)*, Wendel Cox Consultancy, Illinois.

Dovey, K. and Pafka, E. (2014) The urban density assemblage: Modelling multiple measures. *Urban Design International*, **19**, 66–76.

Dubai Electricity and Water Authority (2014) DEWA Sustainability Report 2014. DEWA, Dubai. http://www.dewa.gov.ae/images/DEWA_Sustainability_report_2014.pdf (accessed on 5/07/2015)

Dubai Electricity and Water Authority (2015a) Water statistics 2014. https://www.dewa.gov.ae/aboutus/waterStats2014.aspx (accessed on 5/07/2015)

Dubai Electricity and Water Authority (2015b) Electricity statistics 2014. http://www.dewa.gov.ae/aboutus/electStats2014.aspx (accessed on 5/07/2015)

Dubai Statistics Center (2007) Statistical Year Book 2007. Dubai Statistics Center, Dubai. https://www.dsc.gov.ae/en-us/Publications/Pages/publication-details.aspx?PublicationId=5&year=2007 (accessed on 7/06/2015)

Dubai Statistics Centre (2012) Statistical Yearbook 2012. Dubai Statistics Center, Dubai. https://www.dsc.gov.ae/en-us/Publications/Pages/publication-details.aspx?PublicationId=5&year=2012 (accessed on 7/06/2015)

Dubai Statistics Center (2015) Dubai Population Clock. https://www.dsc.gov.ae/en-us/Pages/default.aspx (accessed on 7/06/2015)

Dyson, T. (2003) HIV/AIDS and urbanization. *Population and Development Review*, **29**, 427–442.

Elvidge, C.D., Tuttle, B.T., Sutton, P.C. *et al.* (2007) Global distribution and density of constructed impervious surfaces. *Sensors*, **7**, 1962–1979.

Er, K.B.H., Innes, J.L., Martin, K. *et al.* (2005) Forest loss with urbanization predicts bird extirpations in Vancouver. *Biological Conservation*, **126**, 410–419.

Fenn, M.E., Baron, J.S., Allen, E.B. *et al.* (2003) Ecological effects of nitrogen deposition in the western United States. *BioScience*, **53**, 404–420.

Fernández-Juricic, E. (2000) Local and regional effects of pedestrians on forest birds in a fragmented landscape. *The Condor*, **102**, 247–255.

Fernández-Juricic, E. and Jokimäki, J. (2001) A habitat island approach to conserving birds in urban landscapes: case studies from southern and northern Europe. *Biodiversity and Conservation*, **10**, 2023–2043.

Forman, R.T.T. (2008) *Urban Regions. Ecology and Planning Beyond the City*, Cambridge University Press, Cambridge.

Forman, R.T.T. (2014) *Urban Ecology: Science of Cities*, Cambridge University Press, Cambridge/New York.

Fuller, R.A., Warren, P.H., and Gaston, K.J. (2007) Daytime noise predicts nocturnal singing in urban robins. *Biology Letters*, **3**, 368–370.

Gago, E.J., Roldan, J., Pacheco-Torres, R. *et al.* (2013) The city and urban heat islands: a review of strategies to mitigate adverse effects. *Renewable Sustainable Energy Review*, **25**, 749–758.

Gammage, G., Stigler, M., Clark-Johnson, S. *et al.* (2011) *Watering the Sun Corridor: Managing Choices in Arizona's Megapolitan Area*. Morrison Institute for Public Policy, Arizona State University, Tempe. https://morrisoninstitute.asu.edu/sites/default/files/content/products/SustPhx_WaterSunCorr.pdf (accessed on 5/07/2015)

Gartland, L. (2008) *Heat Islands: Understanding and Mitigating Heat in Urban Areas*, Earthscan Publications, London.

Gaston, K.J., Bennie, J., Davies, T.W. *et al.* (2013) The ecological impacts of nighttime light pollution: a mechanistic appraisal. *Biological Reviews*, **88**, 912–927.

Gibb, H. and Hochuli, D.F. (2002) Habitat fragmentation in an urban environment: large and small fragments support different arthropod assemblages. *Biological Conservation*, **106**, 91–100.

Graham, A.L. and Cooke, S.J. (2008) The effects of noise disturbance from various recreational boating activities common to inland waters on the cardiac physiology of a freshwater fish, the largemouth bass (*Micropterus salmoides*). *Aquatic Conservation: Marine and Freshwater Ecosystems*, **18**, 1315–1324.

Grimm, N.B., Faeth, S.H., Golubiewski, N.E. *et al.* (2008a) Global change and the ecology of cities. *Science*, **319**, 756–760.

Grimm, N.B., Foster, D., Groffman, P. *et al.* (2008b) The changing landscape: ecosystem responses to urbanization and pollution across climatic and societal gradients. *Frontiers in Ecology and the Environment*, **6**, 264–272.

Guhathakurta, S. and Gober, P. (2007) The impact of the Phoenix urban heat island on residential water use. *Journal of the American Planning Association*, **73**, 317–329.

Hamberg, L., Lehvavirta, S., Malmivaara-Lamsa, M. *et al.* (2008) The effects of habitat edges and trampling on the understorey vegetation in urban forests in Helsinki, Finland. *Applied Vegetation Science*, **11**, 83–98.

Harada, H., Dong, N.T., and Matsui, S. (2008) A measure of provisional and urgent sanitary improvement in developing countries: septic tank performance improvement. *Water Science and Technology*, **58**, 1305–1311.

Harlan, S.L., Brazel, A.J., Jenerette, G.D. *et al.* (2007) In the shade of affluence: the inequitable distribution of the urban heat island. *Research in Social Problems and Public Policy*, **15**, 173–202.

Haviland-Howell, G., Frankel, A.S., Powell, C.M. *et al.* (2007) Recreational boating traffic: a chronic source of anthropogenic noise in the Wilmington, North Carolina intracoastal waterway. *Journal of the Acoustical Society of America*, **122**, 151–160.

Henry, S., Boyle, P., and Lambin, E.F. (2003) Modelling inter-provincial migration in Burkina Faso, West Africa: the role of sociodemographic and environmental factors. *Applied Geography*, **23**, 115–136.

Hettige, H., Mani, M., and Wheeler, D. (1998) *Industrial Pollution in Economic Development: Kuznets Revisited*, The World Bank, Development Research Group, Washington D.C.

Hoornweg, D. and P. Bhada-Tata (eds) (2012) *What a Waste – A Global Review of Solid Waste Management*. World Bank, Washington, DC.

Howard, L. (1833) *Climate of London Deduced From Meteorological Observations*, vol. **1**, Harvey and Darton, London.

Hynes, L.N., McDonnell, M.J., and Williams, N.S.G. (2004) Measuring the success of urban riparian revegetation projects using remnant vegetation as a reference community. *Ecological Management and Restoration*, **5**, 205–209.

Ishan, A.I., Norddin, N.A.M., Malek, N.A. *et al.* (2014) The quality of housing environment and green open space towards quality of life, in *Fostering Ecosphere in the Built Environment, UMRAN2014* (eds A.A. Bakar and N.A. Malek), International Islamic University Malaysia, Kuala Lumpur, pp. 183–198.

Kentula, M.E., Gwin, S.E., and Pierson, S.M. (2004) Tracking changes in wetlands with urbanization: sixteen years of experience in Portland, Oregon, USA. *Wetlands*, **24**, 734–743.

Khan, S. (2014) Sanitation and wastewater technologies in Harappa/Indus Valley Civilization (ca. 2600-1900 BC), in *Evolution of Sanitation and Wastewater Technologies Through the Centuries* (eds A.N. Angelakis and J.B. Rose), IWA Publishing, London, pp. 25–42.

Krivoshto, I.N., Richards, J.R., Albertson, T.E. *et al.* (2008) The toxicity of diesel exhaust: implications for primary care. *Journal of the American Board of Family Medicine*, **21**, 55–62.

Kromroy, K., Ward, K., Castillo, P. *et al.* (2007) Relationships between urbanization and the oak resource of the Minneapolis/St. Paul metropolitan area from 1991 to 1998. *Landscape and Urban Planning*, **80**, 375–385.

Lacoste, E. and Chalmin, P. (2007) *From Waste to Resource: 2006 World Waste Survey*, Economica, Paris.

Lande, R. (1993) Risks of population extinction from demographic and environmental stochasticity and random catastrophes. *The America Naturalist*, **142**, 911–927.

Landsberg, H.E. (1981) *The Urban Climate*, Academic Press, New York.

Leary, E. and McDonnell, M. (2001) Quantifying public open space in metropolitan Melbourne. *Australian Parks and Leisure*, **4**, 34–36.

Lesage, V., Barrette, C., Kingsley, M.C.S. *et al.* (1999) The effects of vessel noise on the vocal behaviour of belugas in the St. Lawrence River Estuary. *Canada Marine Mammal Science*, **15**, 65–84.

Liu, J., Zhan, J., and Deng, X. (2005) Spatio-temporal patterns and driving forces of urban land expansion in China during the economic reform era. *Ambio*, **34**, 450–455.

Löfvenhaft, K., Björn, C., and Ihse, M. (2002) Biotope patterns in urban areas: a conceptual model integrating biodiversity issues in spatial planning. *Landscape and Urban Planning*, **58**, 223–240.

Löfvenhaft, K., Runborg, S., and Sjögren-Gulve, P. (2004) Biotope patterns and amphibian distribution as assessment tools in urban landscape planning. *Landscape and Urban Planning*, **68**, 403–427.

Longcore, T. and Rich, C. (2004) Ecological light pollution. *Frontiers in Ecology and the Environment*, **2**, 191–198.

Luck, G.W., Ricketts, T.H., Daily, G.C. *et al.* (2004) Alleviating spatial conflict between people and biodiversity. *Proceedings of the National Academy of Sciences of the United States of America*, **101**, 182–186.

Luck, M. and Wu, J. (2002) A gradient analysis of urban landscape pattern: a case study from the Phoenix metropolitan region, Arizona, USA. *Landscape Ecology*, **17**, 327–339.

Lyon, L. and R. Driskell (eds) (2011) *The Community in Urban Society*. Waveland Press, Long Grove.

Macionis, J.J. and Parrillo, V.N. (2012) *Cities and Urban Life*, 6th edn, Pearson, Upper Saddle River.

Matsuoka, R.H. and Kaplan, R. (2008) People needs in the urban landscape: Analysis of *Landscape and Urban Planning* contributions. *Landscape and Urban Planning*, **84**, 7–19.

May, R. (1974) *Stability and Complexity in Model Ecosystems*, Princeton University Press, Princeton.

McDonald, R.I., Forman, R.T.T. and Kareiva, P.M. (2010) Open space loss and land inequality in United States' Cities, 1990-2000. *PLoS One* **5**(3): e9509. doi:10.1371/journal.pone.0009509

McGranahan, G., Balk, D., and Anderson, B. (2007) The rising tide: assessing the risk of climate change and human settlements in low elevation coastal zones. *Environment and Urbanization*, **19**, 17–37.

McGranahan, G., Marcotullio, P., Bai, X. *et al.* (2005) Urban systems, in *Ecosystems and Human Wellbeing: Current Status and Trends* (eds R. Hassan, R. Scholes, and N. Ash), Island Press, Washington, DC, pp. 795–825.

McKinney, M.L. (2002) Urbanization, biodiversity, and conservation. *BioScience*, **52**, 883–890.

McMichael, A.J. (2000) The urban environment and health in a world of increasing globalization: issues for developing countries. *Bulletin of the World Health Organization*, **78**, 1117–1127.

Meredith, W. (2005) Mosquito control: balancing public health and the environment in Delaware. *NOAA Coastal Services Magazine*: September/October 2005. http://www.csc.noaa.gov/magazine/2005/05/article2.html (accessed on 18/07/2015)

Morris, D.W. and Kingston, S.R. (2002) Predicting future threats to biodiversity from habitat selection by humans. *Evolutionary Ecology Research*, **4**, 787–810.

Murgui, E. (2007) Effects of seasonality on the species-area relationship: a case study with birds in urban parks. *Global Ecology and Biogeography*, **16**, 319–329.

Nassauer, J.I. (2004) Monitoring the success of metropolitan wetland restorations: cultural sustainability and ecological function. *Wetlands*, **24**, 756–765.

Niemela, J. (1999) Ecology and urban planning. *Biodiversity and Conservation*, **8**, 119–131.

Njeru, J. (2006) The urban political ecology of plastic bag waste problem in Nairobi, Kenya. *Geoforum*, **37**, 1046–1058.

Nowak, J.N., Rowntree, R.A., McPherson, E.G. *et al.* (1996) Measuring and analyzing urban tree cover. *Landscape and Urban Planning*, **36**, 49–57.

Oke, T.R. (1982) The energetic basis of the urban heat island. *Quarterly Journal of the Royal Meteorological Society*, **108**, 1–24.

Ouyang, Z., Xiao, R.B., Schienke, E.W. *et al.* (2008) Beijing urban spatial distribution and resulting impacts on heat islands, in *Landscape Ecological Applications in Man-influenced Areas: Linking Man and Nature Systems* (eds S.K. Hong, N. Nakagoshi, B.J. Fu, and Y. Morimoto), Springer, Dordrecht, pp. 459–478.

Pacione, M. (2005) City profile Dubai. *Cities*, **22**, 255–265.

Pargal, S., Hettige, H., Singh, M. *et al.* (1997) Formal and informal regulation of industrial pollution: comparative evidence from Indonesia and the United States. *The World Bank Economic Review*, **11**, 433–450.

Parés-Ramos, I.K., Gould, W.A. and Aide, T.M (2008) Agricultural abandonment, suburban growth, and forest expansion in Puerto Rico between 1991 and 2000. *Ecology and Society* **13**(2), 1. http://www.ecologyandsociety.org/vol13/iss2/art1/

Parris, K.M., Velik-Lord, M. and North, J.M.A (2009) Frogs call at a higher pitch in traffic noise. *Ecology and Society* **14**(1), 25. http://www.ecologyandsociety.org/vol14/iss1/art25/

Patricelli, G.L. and Blickley, J.L. (2006) Avian communication in urban noise: causes and consequences of vocal adjustment. *The Auk*, **123**, 639–649.

Pauchard, A., Aguayo, M., Pena, E. *et al.* (2006) Multiple effects of urbanization on the biodiversity of developing countries: the case of a fast-growing metropolitan area (Concepcion, Chile). *Biological Conservation*, **127**, 272–281.

Pearce, F. (1996) How big can cities get? *New Scientist*, **190**, 10.

Rebele, F. (1994) Urban ecology and special features of urban ecosystems. *Global Ecology and Biogeography Letters*, **4**, 173–187.

Relyea, R.A. (2005) The lethal impacts of roundup on aquatic and terrestrial amphibians. *Ecological Applications*, **15**, 1118–1124.

Riley, S.P.D., Busteed, G.T., Kats, L.B. *et al.* (2005) Effect of urbanization on the distribution and abundance of amphibian and invasive species in southern California streams. *Conservation Biology*, **19**, 1894–1907.

Rosenzweig, M.L. (2003) *Win-Win Ecology: How the Earth's Species Can Survive in the Midst of Human Enterprise*, Oxford University Press, Oxford.

Sattler, T., Borcard, D., Arlettaz, R. *et al.* (2010) Spider, bee, and bird communities in cities are shaped by environmental control and high stochasticity. *Ecology*, **91**, 3343–3353.

Scalenghe, R. and Marsan, F.A. (2008) The anthropogenic sealing of soils in urban areas. *Landscape and Urban Planning*, **90**, 1–10.

Seto, K.C., Güneralp, B., and Hutyra, L.R. (2012) Global forecasts of urban expansion to 2030 and direct impacts on biodiversity and carbon pools. *Proceedings of the National Academy of Sciences*, **40**, 16083–16088.

Shochat, E. (2004) Credit or debit? Resource input changes population dynamics of city-slicker birds. *Oikos*, **106**, 622–626.

Shrestha, M.K., York, A.M., Boone, C.G. *et al.* (2012) Land fragmentation due to rapid urbanization in the Phoenix Metropolitan Area: analysing the spatiotemporal patterns and drivers. *Applied Geography*, **32**, 522–531.

Shukuroglou, P. and McCarthy, M.A. (2006) Modelling the occurrence of rainbow lorikeets (*Trichoglossus haematodus*) in Melbourne. *Austral Ecology*, **31**, 240–253.

Slabbekoorn, H., Yeh, P., and Hunt, K. (2007) Sound transmission and song divergence: a comparison of urban and forest acoustics. *The Condor*, **109**, 67–78.

Spinks, P.Q., Pauly, G.B., Crayon, J.J. *et al.* (2003) Survival of the western pond turtle (*Emys marmorata*) in an urban California environment. *Biological Conservation*, **113**, 257–267.

Sturman, A.P. and Tapper, N.J. (2006) *Weather and Climate in Australia and New Zealand*, Oxford University Press, Oxford.

Suren, A.M., Riis, T., Biggs, B.J.F. *et al.* (2005) Assessing the effectiveness of enhancement activities in urban streams: I. Habitat responses. *River Research and Applications*, **21**, 381–401.

Tait, C., Daniels, C.B., and Hill, R.S. (2005) The urban ark: The historical evolution of the plant community, in *Adelaide. Nature of a City: The Ecology of a Dynamic City from 1836 to 2036* (eds C.B. Daniels and C. Tait), BioCity, Adelaide, pp. 87–110.

Tan, J., Zheng, Y., Tang, X. *et al.* (2010) The urban heat island and its impacts on heat waves and human health in Shanghai. *International Journal of Biometeorology*, **54**, 75–84.

Torok, S.J., Morris, C.J.G., Skinner, C. *et al.* (2001) Urban heat island features of southeast Australian towns. *Australian Meteorological Magazine*, **50**, 1–13.

UAE Interact (2007) *UAE Yearbook 2007*. http://www.uaeyearbook.com/Yearbooks/2007/ENG/ (accessed on 24/02/2014).

UNEP (2005) *Selection, Design and Implementation of Economic Instruments in the Solid Waste Management Sector in Kenya: The Case of Plastic Bags*. UNEP Division of Technology, Industry and Economics, Paris. http://www.unep.ch/etb/publications/econinst/kenya.pdf (accessed on 18/07/2015).

UNFPA (2007) *State of World Population 2007. Unleashing the Potential of Urban Growth*, United Nations Population Fund, New York.

US Energy Information Administration (2015) Arizona State Profile and Energy Estimates. http://www.eia.gov/state/?sid=AZ (accessed on 7/06/2015)

US Environmental Protection Agency (2014) *Municipal Solid Waste Generation, Recycling, and Disposal in the United States: Facts and Figures for 2012*. http://www.epa.gov/osw/nonhaz/municipal/pubs/2012_msw_fs.pdf (accessed on 7/06/2015)

Van de Poel, E., O'Donnell, O., and Van Doorslaer, E. (2009) What explains the rural-urban gap in infant mortality: household or community characteristics? *Demography*, **46**, 827–850.

Walsh, C.J., Papas, P.J., Crowther, D. *et al.* (2004) Storm water drainage pipes as a threat to a stream-dwelling amphipod of conservation significance, *Austrogammarus australis*, in south-eastern Australia. *Biodiversity and Conservation*, **13**, 781–793.

Walsh, C.J., Roy, A.H., Feminella, J.W. *et al.* (2005) The urban stream syndrome: current knowledge and the search for a cure. *Journal of the North American Benthological Society*, **24**, 706–723.

Warren, P.S., Katti, M., Ermann, M. *et al.* (2006) Urban bioacoustics: it's not just noise. *Animal Behaviour*, **71**, 491–502.

WHO (2014a) Ambient air pollution database, May 2014. http://www.who.int/phe/health_topics/outdoorair/databases/cities/en/ (accessed on 8/06/2015).

WHO (2014b) Ambient (outdoor) air quality and health. Fact Sheet No. 313, updated March 2014. http://www.who.int/mediacentre/factsheets/fs313/en/ (accessed on 18/07/2015).

WHO and UN-Habitat (2010) *Hidden Cities: Unmasking and Overcoming Health Inequities in Urban Settings*, WHO Press, Geneva.

Williams, N.S.G., McDonnell, M.J., Phelan, G.K. *et al.* (2006) Range expansion due to urbanization: Increased food resources attract Grey-headed Flying-foxes (*Pteropus poliocephalus*) to Melbourne. *Austral Ecology*, **31**, 190–198.

WMO (2015a) World Weather information Service: Dubai, United Arab Emirates. http://worldweather.wmo.int/en/city.html?cityId=1190 (accessed on 7/06/2015)

WMO (2015b) World Weather information Service: Phoenix, Arizona. http://worldweather.wmo.int/en/city.html?cityId=806 (accessed on 7/06/2015)

World Bank (2013) *Global Monitoring Report 2013: Rural-Urban Dynamics and the Millennium Development Goals*, World Bank, Washington, DC.

WWF (2014) *Living Planet Report 2014: Species and Spaces*, People and Places, WWF, Gland, Switzerland.

Xiao, R., Ouyang, Z., Zheng, H. *et al.* (2007) Spatial pattern of impervious surfaces and their impacts on land surface temperature in Beijing, China. *Journal of Environmental Sciences*, **19**, 250–256.

CHAPTER 3

Population- and species-level responses to urbanization

3.1 Introduction

The biophysical processes of urbanization can have wide-ranging effects on populations and species by altering the quantity, quality, temporal and spatial arrangement of resources that microbes, fungi, plants and animals depend on for survival, such as shelter, nest sites, food, water, sunlight and nutrients (e.g., Riley et al. 2003; Shochat et al. 2006; Harper et al. 2008; McDonald et al. 2008). Cities also experience many biological introductions and invasions, with some species dispersing into urban environments and many others being deliberately or inadvertently introduced by humans (Rebele 1994; Lambdon et al. 2008; Williams et al. 2009). In addition, the high density of humans in urban areas increases the probability that people will disturb other organisms and their habitat. Common disturbances include trampling of plants and animals, harvesting of fungi and plants, hunting animals for food, and altering the behavioural patterns of animals sensitive to the presence of humans (Fernández-Juricic 2000; Milner-Gulland et al. 2003; Murison et al. 2007; Peres and Palacious 2007; Florgård 2009). These processes (the primary and secondary biophysical processes of urbanization, biological introductions and human disturbance) combine to affect the abundance of non-human species in urban environments. They do this in part by altering the interactions within and between species, including competition for resources, predation, cannibalism, parasitism, symbiosis and mutualism, as well as more complex interactions such as predator-mediated competition.

Population growth and decline are a function of four processes or vital rates: births, deaths, immigration and emigration. Therefore, a logical way to think about how populations (and the species they belong to) respond to urbanization is to consider how urbanization impacts upon vital rates, ultimately leading to population growth or decline. Figure 3.1 shows a conceptual model of the important mechanisms and pathways through which the environmental changes associated with urbanization can impact on non-human populations. Disturbance by humans, changes in the availability of resources, and changes to intra- and inter-specific interactions can affect the survival of juveniles and adults, the production of offspring, and the number of individuals that disperse into or

Ecology of Urban Environments, First Edition. Kirsten M. Parris.

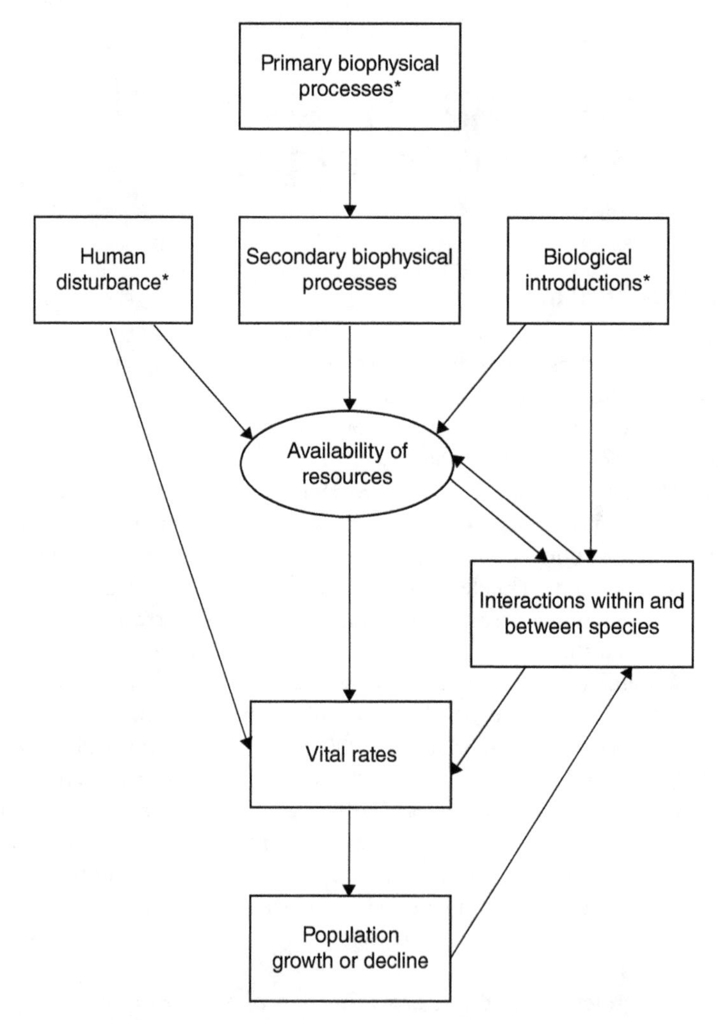

Figure 3.1 A conceptual model of the processes associated with urbanization, and the important mechanisms and pathways through which these can affect populations of non-human species. *Processes that are strongly influenced by human preferences.

out of urban populations. Environmental and demographic stochasticity and catastrophic events (as defined in Chapter 2) may also influence vital rates in urban environments. The subsequent growth or decline of populations can then further alter interactions within and between species, establishing a feedback loop through which populations of different species continue to increase or decrease, sometimes to the point of local extinction (Figure 3.1). For example, the Xerces blue (*Glaucopsyche xerces*) is thought to be the first American butterfly to become extinct due to urban development in San Francisco, California (Pyle 1995).

In Figure 3.1, I have indicated with an asterisk three groups of processes associated with the construction and expansion of cities that are strongly influenced by human preferences; the primary biophysical processes of urbanization, human disturbance and biological introductions. Aesthetics, cultural preferences, population pressure, socio-economics and the speed of urban development can influence the physical form of cities (e.g., how much vegetation is cleared, how much open space and permeable surface remain, the density of roads and housing, and whether natural features such as wetlands are retained, modified or destroyed; Sharpe et al. 1986; Montgomery 1998; Clark et al. 2002; Kentula et al. 2004). Human preferences also influence patterns of activity in different parts of a city, such as which places are visited, when, by how many people, and what they do there (Burgess et al. 1988; Storper and Manville 2006).

Nature enthusiasts may be most likely to walk though parks or remnant patches of native vegetation in spring and summer, potentially trampling plants or disturbing breeding birds. Urban-dwellers may harvest wild plants and animals for food or to sell in markets, or collect standing and fallen timber for fuel (Maharjan 1998; Asfaw and Tadesse 2001). In addition, human preferences dictate the types of non-native organisms that people deliberately introduce into urban habitats (e.g., Hope et al. 2003; Muerk et al. 2009). A wide range of cultivated garden plants, including trees, shrubs, flowers, herbs and vegetables, domestic animals such as cats, dogs, chickens and pigs, and ornamental fish such as goldfish and koi carp have been introduced to cities all over the world. Domestic cats and dogs are almost ubiquitous urban dwellers, while proteas from South Africa or rhododendrons from Japan can be found in city gardens in Australia, Europe and North America. Below, I examine in more detail how non-human populations and species respond to these aspects of urbanization.

3.2 Responses to the secondary biophysical processes of urbanization

3.2.1 Habitat loss, fragmentation and isolation

Many microbes, fungi, plants and animals die when their habitat is lost following urbanization, although some microbial and fungal species may persist in the soil and individuals of certain animal taxa may be able to disperse successfully to neighbouring areas (Czech and Krausman 1997; Czech et al. 2000; How and Dell 2000; Faulkner 2004; Newbound 2008; Newbound et al. 2010). Thus, the impacts of habitat loss on populations are usually obvious and immediate. In contrast, the impacts of habitat fragmentation (the dissection of remaining habitat into small areas or patches) and habitat isolation (the physical separation of habitat patches by areas of "non-habitat") can be more subtle and take many generations to manifest (Table 3.1). Metapopulation theory (Levins 1969) suggests that patches of habitat surrounded by an altered landscape may be analogous to oceanic islands. The theory predicts that small patches are less likely to support a population of a

Table 3.1 A summary of the main effects of habitat loss, fragmentation and isolation on populations and species in urban environments.

Impacts of habitat loss, fragmentation and isolation on populations and species
Habitat loss causes:
Reduced survival of individuals
Emigration of individuals
Local extinction of populations
Habitat fragmentation leads to:
Smaller areas of habitat that support smaller populations
A higher probability of local extinction
Reduced species richness
Habitat isolation leads to:
Reduced movement of individuals between areas of habitat
A lower probability of colonization or re-colonization following local extinction
Lower rates of pollination, disrupted dispersal of seeds and spores
Reduced gene flow between areas of habitat and a loss of genetic diversity

given species than large patches because of a higher probability of local extinction, while isolated patches are less likely to support a population than patches close to other suitable habitat because of a lower probability of colonization (immigration) from another patch (Hanski 1994, 1998; Figure 3.2). Consequently, populations in small and/or isolated patches of habitat surrounded by a sea of urban development may continue to be lost over time, an idea known as the extinction debt (Tilman et al. 1994; McCarthy et al. 1997). Because of the lag time between landscape change and population-level responses, there could still be a substantial extinction debt to pay for past fragmentation and isolation of natural habitats in cities around the world (Hanski and Ovaskainen 2002; Hahs et al. 2009; Kuussaari et al. 2009). The current presence of populations in habitat fragments may therefore not be a good indication of their likely persistence in the future, particularly for long-lived species (Spinks et al. 2003).

For this metapopulation (or fragmentation) model to hold, the remnant patches of habitat need to be surrounded by hostile "non-habitat" that does not provide resources such as food or shelter, and that is difficult or costly to traverse (Bunnell 1999; Lindenmayer and Franklin 2002). These conditions may be met for species that require particular resources or conditions that do not occur in urban areas outside designated patches of habitat (e.g., Heard et al. 2013), and/or for species that cannot successfully disperse between habitat patches. They are also more likely to be met where habitat patches are surrounded by very dense urban development with a high cover of impermeable surfaces (i.e., areas that provide few resources for any species). Drinnan (2005) found that large woodland reserves in southern Sydney supported populations of more species of fungi, plants, birds and frogs than small reserves, while more isolated reserves supported populations of fewer species than those connected by habitat corridors or in close proximity

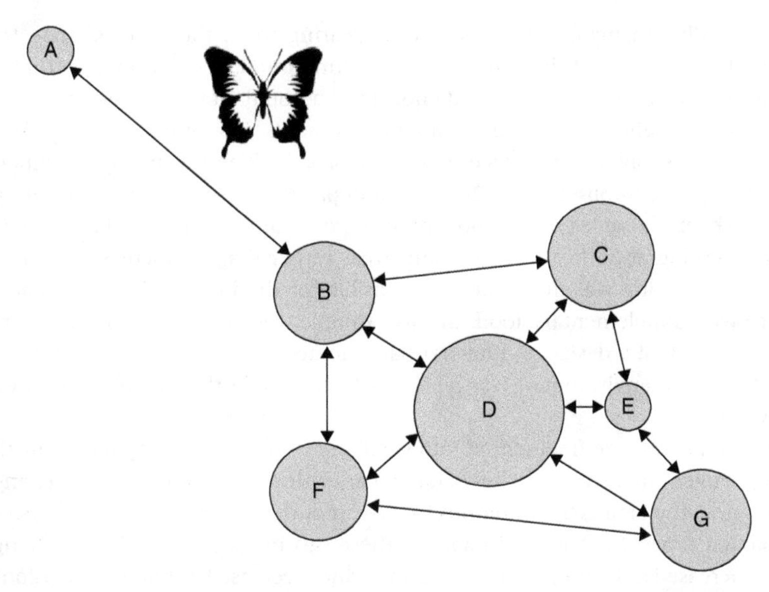

Figure 3.2 A hypothetical habitat network for a butterfly comprised of seven habitat patches of varying size and proximity to each other (circles); the arrows show potential pathways for dispersal between patches. Metapopulation theory predicts that small and/or isolated patches are less likely to support a population than large and/or connected patches. In this case, patch A (small and isolated) is least likely to support a population, while patch D (large and well-connected) is most likely.

to other large reserves. In an early study of urban amphibians in Sussex, England, frogs, toads and newts were less likely to occur in garden ponds as the surrounding density of urban development increased (Beebee 1979).

A useful alternative to the binary classification of habitat and non-habitat in urban areas is the continuum model, which allows for gradual changes in the resources required by different species across space (Fischer and Lindenmayer 2006). This may be a more appropriate model for a range of species inhabiting urban and non-urban environments alike. For example, some species of native possum are abundant in the cities of eastern Australia, and can reach much higher densities in urban than non-urban habitats (Harper 2005). A study of common ringtail and brushtail possums in patches of remnant woodland in Melbourne found that their abundance increased with the availability of resources both inside the patches and in the surrounding urban area (Harper et al. 2008). Higher densities of each possum species occurred in woodland patches with more potential den sites (hollow-bearing trees) and a higher cover of trees and shrubs in a 100-m buffer surrounding the remnant. Vegetation outside the woodland patches was largely comprised of exotic garden plants that are more palatable and nutritious for the possums than the indigenous vegetation that has evolved chemical defences against browsing animals.

In woodland patches with few hollow-bearing trees, the cover of food trees in the buffer area had little impact on possum abundance – it appears that the possums need ample den sites to demonstrate a population-level response to the increased availability of food in urban environments. In a resource-manipulation experiment, supplementary feeding with apples doubled the average number of brushtail possums observed in 20 woodland patches in only six months (Harper 2005). Given the speed of the population response and the timing of the breeding season, immigration from surrounding areas rather than an increase in survival rates or fecundity was most likely responsible for the increase in possum abundance with supplementary food. In this example, the possums could easily cross the boundary of a designated habitat patch to access resources in adjacent areas and the urban environment was relatively benign (suburban developments with leafy gardens).

When habitats are fragmented into small areas, they are susceptible to further changes over time. Small patches of habitat are vulnerable to edge effects – changes in physical and biological conditions that occur at the boundary of two ecosystems or habitat types. In terrestrial habitats, these can include microclimatic changes (e.g., increased radiation, temperature and wind, decreased humidity), invasion by exotic species, increased pressure from herbivores and predators, influxes of pollutants and increased disturbance by humans (Morgan 1998; Beale and Monaghan 2004; Zhou et al. 2004; Grimm et al. 2008). Disturbed and/or fragmented aquatic habitats in urban environments also suffer from edge effects, such as invasion by exotic species and increased predation pressure, as well as altered hydrological and thermal regimes (Temple 1987; King and Buckney 2000; Trombulak and Frissell 2000; Konrad and Booth 2005). Many of these changes then alter the availability of resources for different species, interactions between species, and/or have direct effects on vital rates in small habitat patches. Certain plant species may have an increased competitive advantage in sunnier conditions at the edge of a habitat patch (Hamberg et al. 2008). In contrast, a higher density of avian predators in edge habitats can increase predation pressure on the nests of other bird species, decreasing breeding success (Gates and Gysel 1978; Batary and Baldi 2004). Disturbance by humans – including collection of firewood and the clearing or trampling of vegetation – can reduce habitat complexity in small remnants, removing structural components that provide refuges for invertebrates and small vertebrates, such as native rodents (Soulé et al. 1992; Sauvajot et al. 1998).

The high density of roads in urban areas represents a substantial barrier to movement for many terrestrial animals, including arthropods, amphibians, reptiles and mammals, effectively isolating populations that remain in patches of habitat within cities (e.g., Gibbs and Shriver 2002; Marchand and Litvaitis 2004; Seiler et al. 2004; Vandergast et al. 2007; Noel et al. 2007; Hale et al. 2013). Daily movements, seasonal migrations and longer-distance dispersal are curtailed or prevented when animals must cross roads to reach their destination. Hels and Buchwald (2001) demonstrated that frogs and newts attempting to cross roads are likely to suffer very high mortality rates. While they may be able to cross quieter

Figure 3.3 Busy highways can act as a barrier to dispersal for the coyote *Canis latrans*. Photograph by Republica. https://pixabay.com/en/coyote-penn-dixie-hamburg-new-1820/. Used under CC0 1.0 https://creativecommons.org/publicdomain/zero/1.0/deed.en.

roads safely, as traffic volumes increase the probability of a successful crossing decreases sharply. For example, an individual common newt *Triturus vulgaris* or crested newt *T. cristatus* has <20% chance of safely crossing a road carrying 10,000 vehicles per day (Hels and Buchwald 2001). The probability that an amphibian could safely cross multiple roads when moving through a densely-roaded urban landscape is close to zero. Road traffic is the largest single cause of death for badgers *Meles meles* in Britain (Clarke et al. 1998), and busy highways can also act as a substantial barrier to dispersal for larger mammals, such as coyotes *Canis latrans* (Figure 3.3) and bobcats *Lynx rufus* in the USA (Riley et al. 2006).

Disruption of animal movement by roads and other urban infrastructure may impact the persistence of fungal and plant populations that depend on certain invertebrates or vertebrates for pollination or the dispersal of their spores and seeds (Bhattacharya et al. 2003; Close et al. 2006; Newbound et al. 2010; Bates et al. 2011). Roads act as barriers to the safe movement of many insects, including beetles, bumblebees, crickets, dragonflies and butterflies (reviewed in Muñoz et al. 2015). Butterflies are frequently killed by vehicles travelling on roads; the number of individuals and species killed generally increases with traffic volume and the width of the road (McKenna et al. 2001; Rao and Girish 2007; Skórka et al. 2013). Certain characteristics or traits of butterflies are also correlated with higher rates of mortality on roads; small-bodied and low-flying butterflies suffer higher rates of mortality (Rao and Girish 2007; Skórka et al. 2013), while mobile species that move frequently and over long distances may be killed at a higher rate than sedentary species (De la Puente et al. 2008). An experiment in Boston, Massachusetts found that bumblebees (*Bombus impatiens* and *B. affinis*) could cross roads or railway lines to reach suitable foraging habitat on the other side, but rarely did so because of innate site fidelity – it was hypothesized that because of the intervening road, the bumblebees perceived this habitat as a separate site (Bhattacharya et al. 2003). These human-built features may therefore restrict bumblebee movement and act to fragment plant populations in urban environments.

The isolation of populations by urban development can be observed in their genetic structure; that is, the genetic differentiation between populations. Reduced gene flow between isolated populations leads to increased genetic structure (differentiation between populations), and in certain circumstances, reduced genetic diversity. In extreme cases, low genetic diversity can result in inbreeding depression, which in turn reduces the survival and fecundity of any offspring produced (Reed and Frankham 2003). Inbreeding is known to adversely affect many wild populations (Hedrick and Kalinowski 2000; Keller and Waller 2002). Studies across a range of taxonomic groups, including insects, amphibians and mammals, have demonstrated a relationship between habitat fragmentation in urban environments and reduced genetic diversity (e.g., Hitchings and Beebee 1998; Robinson and Marks 2001; McClenaghan and Truesdale 2002; Vandergast et al. 2007). Populations that are both small and isolated are also vulnerable to genetic bottlenecks, which result from short-term but pronounced declines in effective population size, and genetic drift, in which certain alleles are not transmitted to the next generation by chance alone (Frankham et al. 2004). Low genetic diversity can also limit the ability of populations to evolve in response to environmental changes or novel predators and pathogens (Frankham et al. 2002).

3.2.2 Climatic changes

Changes in temperature and moisture regimes following urbanization can influence the timing of life cycle events (phenology) of fungi, plants and animals in urban areas. In temperate climates, the most common patterns observed include earlier flowering, leafing and fruiting of plants in warmer urban areas (Roetzer et al. 2000; White et al. 2002; Menzel 2003) and earlier reproduction in spring-breeding animals (e.g., Luniak 2004; Partecke et al. 2005; Chamberlain et al. 2009). Differential responses of species to warming can have important effects on trophic systems if the peak availability of prey and the peak food requirements of predators no longer coincide (Durant et al. 2007; Both et al. 2008). Artificial watering of parks and gardens may lengthen and reduce variability in the flowering and fruiting season of plants in urban environments, thereby extending the period of the year in which the flowers and fruits are available as a food resource for other species. In the hot, dry climate of Phoenix, Arizona, the urban heat-island effect has increased summer heat stress on plants, including trees, shrubs and cool-season grasses, while decreasing winter cold stress on cold-sensitive species, such as cacti, succulents and warm-season grasses (Baker et al. 2002). It has also extended the annual period of arthropod activity in the city (the thermal window between 15 and 38 °C) by approximately one month since 1948.

Urban climate change may attract immigrants to cities, such as the grey-headed flying-fox *Pteropus poliocephalus* in Melbourne, Australia (Parris and Hazell 2005). Traditionally considered a warm-temperate to sub-tropical species, a few individuals of *P. poliocephalus* established a permanent (year-round) colony or camp in Melbourne in 1986, and by 2003 the population had grown to approximately 30,000 (van der Ree et al. 2006). On the basis of long-term data, Melbourne does not fall within the climatic range of other *P. poliocephalus* camp sites in Australia (Parris and Hazell 2005). However, due to a general warming trend observed across Australia (Torok and Nicholls 1996) and a strengthening of the urban heat-island effect (Torok et al. 2001; Lenten and Moosa 2003), temperatures in central Melbourne have been increasing since the 1950s. In addition, artificial watering of parks and gardens may contribute the equivalent of 590 mm (95% CI: 450–720 mm) of extra rainfall per year (Parris and Hazell 2005). Human activities appear to have increased temperatures and effective precipitation in central Melbourne, creating a more suitable climate for camps of the grey-headed flying-fox. Conversely, climatic changes in urban areas may lead to the local extinction of populations that lose their climatic niche. Extensive urban development near Basel, Switzerland is implicated in the local extinction of eight populations of the land snail *Arianta arbustorum* (Baur and Baur 1993). The hatching success of the snail's eggs decreases with exposure to temperatures ≥22 °C, demonstrating one mechanism by which elevated temperatures can affect vital rates and, ultimately, the persistence of species in urban environments.

3.2.3 Altered hydrological regimes

Changes to hydrological regimes in urban streams, ponds and wetlands impact upon a wide range of aquatic organisms, including algae, invertebrates, fish and amphibians (Walsh et al. 2005; Hamer and McDonnell 2008). The flashiness of flow in urban streams and the resulting changes to stream morphology and water quality alter the availability of resources, such as food, shelter, attachment and breeding sites; modify interactions between taxa; and ultimately impact upon vital rates. High flows can also flush individual animals downstream and scour algae and plants from the stream bed. Consequently, urban streams tend to support populations of species that are tolerant of disturbance, while many other species are excluded (Walsh et al. 2005). Hydraulic modelling of the River Tame, Birmingham, compared patterns of flow velocity with the maximum sustainable swimming speed (MSSS) of three species of fish – chub, dace and roach (Booker 2003). In most of the available habitat, the river velocity exceeded the MSSS of these fish 16 times in one year, indicating that high flows were likely to exclude them from the river.

Urbanization can change the average hydroperiod of ponds and wetlands in cities (i.e., the period of time that they contain water) by diverting water into or away from them, deepening existing ponds to make them more permanent, and destroying ephemeral wetlands to make way for urban development (Hamer and McDonnell 2008). Longer hydroperiods favour the persistence both of fish and of amphibians with long larval life spans that have evolved to coexist with fish and invertebrate predators in permanent ponds. In contrast, shorter hydroperiods favour amphibians with short larval life spans that are adapted to ephemeral ponds with few predators (Hamer and McDonnell 2008). The hydroperiod of wetlands increased with urbanization in central Pennsylvania, USA (Rubbo and Kiesecker 2005). These urban wetlands tended to contain fish and supported fewer species of amphibian larvae than rural wetlands in the study area. Wood frogs *Rana sylvatica,* spotted salamanders *Ambystoma maculatum* and Jefferson's salamanders *A. jeffersonianum* were rarely found in urban wetlands; as well as responding to the loss of forested habitat in urban areas, their larvae appear to be particularly vulnerable to fish predation (Rubbo and Kiesecker 2005). In this case, a secondary biophysical change associated with urbanization (increasing hydroperiod) increases the availability of a key resource for fish and certain species of amphibians (permanent water), which in turn has a positive effect on their survival and reproduction. As their populations flourish, interactions between these and other species in the ponds are altered. A higher density of fish increases the rate of predation on the larvae of vulnerable amphibians (those adapted to ephemeral ponds without fish), leading to a decline in their populations (Figure 3.1). Amphibian larvae adapted to permanent water may also out-compete those adapted to ephemeral habitats for food (Wellborn et al. 1996).

3.2.4 Pollution of air, water and soil

Urban pollution and waste impact on populations and species in both terrestrial and aquatic habitats. In many cases, they are a source of toxic substances, such as heavy metals and organic compounds, that increase the morbidity and mortality of organisms (Snodgrass et al. 2008; Foster et al. 2014; Morrissey 2014); but in others they provide resources to be exploited, such as nutrients, food and shelter (Gehrt 2004; Grimm et al. 2008). Populations of species that are sensitive to pollutants tend to decline as a consequence of lower survival rates, reduced breeding success, and/or an inferior ability to compete with other species in polluted environments, while species that are tolerant of pollution and waste may thrive (Agneta and Burton 1990; Walsh et al. 2007; Yule et al. 2015). The response of aquatic macroinvertebrates to pollution is well understood, and the presence and abundance of different taxa in the assemblage are considered to be reliable indicators of water quality (Resh and Jackson 1993; Johnson et al. 2013).

A recent study of urban pollution in a section of the Borkena River, Ethiopia, between the towns of Dessie and Kombolcha found the macroinvertebrate assemblage to be substantially altered from that at a reference site upstream

(Beyene et al. 2009). Both the overall abundance of invertebrates and the number of species present decreased with increasing pollution of the water, leading to communities dominated by midges (family Chironomidae) and sludge worms (Tubificidae). Sampling sites impacted by urban pollution had high levels of nitrogen and phosphorus and low levels of dissolved oxygen. At two sites, the five-day biochemical oxygen demand (BOD_5 the amount of dissolved oxygen required by micro-organisms to decompose the organic matter in a water sample over five days) exceeded 1100 mg L^{-1}. For comparison, the recommended BOD_5 level for the protection of cultured fish, molluscs and crustaceans from physico-chemical stress is <15 mg L^{-1} (ANZECC 2000). Dessie and Kombolcha have no formal sewerage system or sewage treatment plant and industrial and municipal waste is discharged directly into the river. The residents of these towns also use untreated river water for drinking and bathing (Beyene et al. 2009), with obvious risks to public health.

Many epiphytic algae, lichens and bryophytes are sensitive to air pollution, particularly sulfur and nitrous oxides (Gilbert 1969; Hawksworth 1970; Leblanc and DeSloover 1970; Puckett et al. 1973; Giordani 2007; Gadsdon et al. 2010). The loss of sensitive taxa from trees in urban areas can have important secondary effects on invertebrates that rely on epiphytes for food and/or shelter. In a study conducted in Newcastle upon Tyne, UK in the 1960s, Gilbert (1971) found that increasing air pollution towards the centre of the city (where SO_2 levels reached an annual average of 224 µg m^{-3}) was correlated with a decline in the cover and diversity of algae and lichens on ash trees. The abundance and species richness of psocids (bark lice) that feed on these algae and lichen also declined as air pollution increased (Figure 3.4). However, improvements in air quality across Europe over recent decades may be helping urban lichen communities to recover (Lättman et al. 2014). For example, the mean annual atmospheric SO_2 concentration in Tampere, Finland, dropped from 160 µg m^{-3} in 1973 to 2 µg m^{-3} in 1999, which coincided with an increase in the species richness and percent cover of epiphytic lichens on the trunks of the common linden tree *Tilia* x *vulgaris* (Ranta 2001). Across 25 study sites, the mean species richness increased from 0.7 to 7.6 species in the period 1980–2000, while the average cover of lichens increased from 0.06 to 10.9%.

Household garbage (municipal waste) provides an important food resource for many city-dwelling animals, including lizards, mice, rats, cats, foxes, coyotes, baboons and bears (Banks et al. 2003; Beckmann and Berger 2003; Contesse et al. 2004; Fox 2006; Powell and Henderson 2008). Unlike many natural food sources, garbage is available throughout the year, has a highly clumped distribution and is constantly replenished (Beckmann and Berger 2003). This resource increases the survival and fecundity and, ultimately, the population sizes of species that utilize it (Robbins et al. 2004; McKinney 2002). Interestingly, animals that eat garbage tend to be active for fewer hours per day than those that rely on natural sources of food, because they need to spend less time foraging (Banks et al. 2003; Beckmann and Berger 2003). This can contribute to health problems, such as

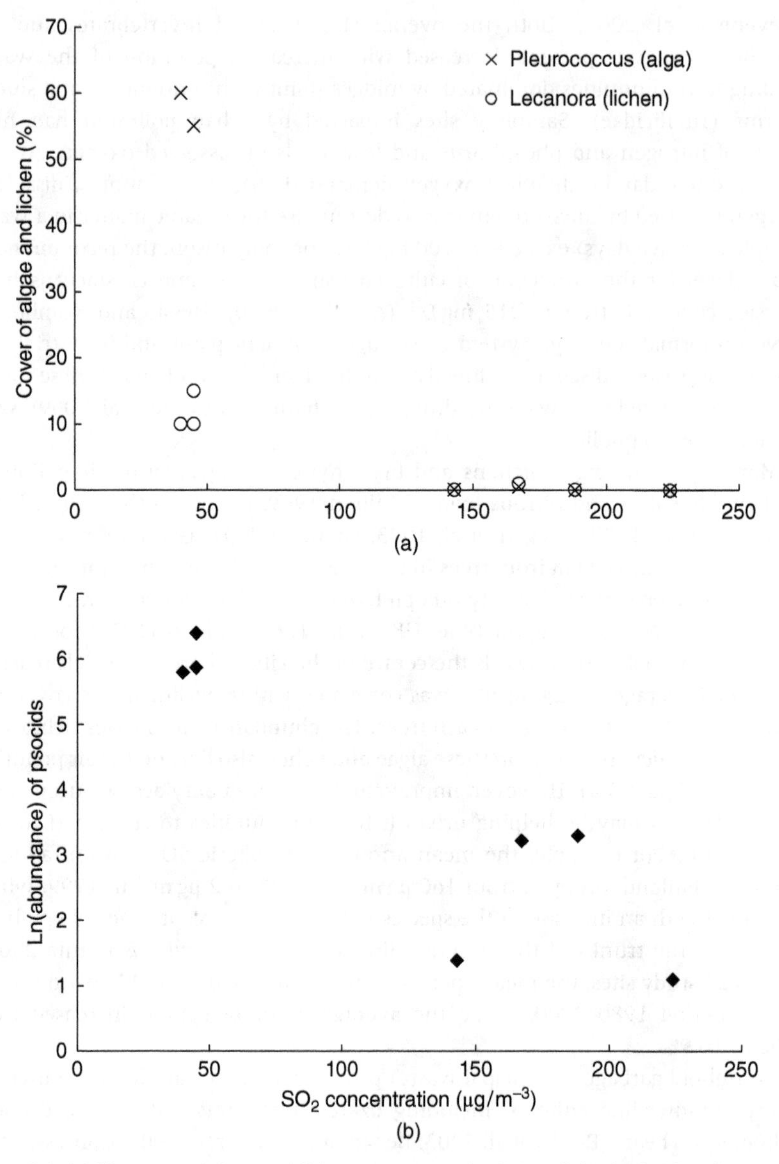

Figure 3.4 (a) The cover of algae and lichen on tree trunks (%) and (b) the ln(abundance) of psocids (bark lice; multiple species represented) as a function of the annual average atmospheric SO_2 concentration in Newcastle upon Tyne, UK; circles = % cover of the alga *Pleurococcus naegeli*, crosses = % cover of the lichen *Lecanora conizaeoides*. Data from Gilbert (1971).

obesity and insulin resistance, as observed in garbage-eating baboons in Kenya (Banks et al. 2003).

3.2.5 Altered noise and light regimes

Communication is the basis of all social relationships between animals. Insects, fish, frogs, birds and mammals use acoustic (sound) signals for a range of social purposes, including begging for food, attracting and bonding with mates, defending territories, maintaining contact with a group, and warning of danger from approaching predators. Noise from road traffic (in terrestrial habitats) or recreational boating and commercial shipping (in aquatic habitats) can hinder acoustic communication by reducing the distance over which a signal can be detected (Warren et al. 2006; Bee and Swanson 2007; Vasconcelos et al. 2007; Box 3.1). This is known as acoustic interference or masking. The noise of passing vehicles can also startle animals, triggering a physiological stress response and changing animal behaviour in a variety of ways (Singer 1978; Sun and Narins 2005; Samuel et al. 2005; Kight and Swaddle 2011; Shannon et al. 2014; Parris 2015). A study of greater sage-grouse *Centrocercus urophasianus* in Wyoming, USA found that male birds avoided lekking sites (places they call from to attract females for mating) with speakers playing recorded traffic noise (Blickley et al 2012a). Male birds that did use the noisy lekking sites had higher levels of stress hormones (glucocorticoids) than males at quiet sites (Blickley et al. 2012b). High levels of background noise also increase watchfulness (or vigilance behaviour) in animals – because they cannot hear predators approaching, prey species spend more time watching for predators and less time foraging for food (Barber et al. 2010). Noise from road construction or road traffic can even cause temporary or permanent hearing loss in animals, including fish, reptiles and marine mammals (Brattstrom and Bondello 1983; Erbe 2002; McCauley et al. 2003; Popper and Hastings 2009; Slabbekoorn et al. 2010).

Box 3.1 Boat Noise and Acoustic Communication in the Lusitanian Toadfish

The Lusitanian toadfish *Halobatrachus didactylus* occurs in estuaries and near-coastal areas of the Mediterranean Sea and eastern Atlantic Ocean. During the breeding season, male toadfish establish territories in shallow water where they build a nest under rocks. In a similar way to male frogs, individual males establish territories close to each other and then call to attract females for mating using a long, low-pitched tonal signal known as a boat whistle (Dos Santos et al. 2000; Vasconcelos et al. 2007). This signal also has a territorial function, and is used to keep intruders out during male-to-male encounters (Vasconcelos et al. 2010). The geographic distribution of the Lusitanian toadfish coincides with busy shipping channels and there is concern that low-frequency underwater noise from ships may be interfering with its acoustic communication. A laboratory study demonstrated that noise from ferry boats raises the auditory thresholds of the toadfish considerably, reducing the distance over which their mating and territorial signals can be heard (Vasconcelos et al. 2007). The characteristics of boat whistles vary between individual males, providing scope for mate choice by females (Amorim and Vasconcelos, 2008). The noise of boat traffic could mask these signal differences and potentially compromise mate selection in this species (Vasconcelos et al. 2007). As has been observed in many species of frogs, a higher calling rate and longer time spent calling are correlated with greater reproductive success in the Lusitanian toadfish (Vasconcelos et al. 2012). A higher calling rate is also an indicator of larger body size and better body condition.

Songbirds show a range of behavioural responses to urban noise, including singing at a higher frequency (pitch) to reduce acoustic interference from the low-frequency noise (a frequency shift), singing more loudly (an amplitude shift), and changing diurnal singing patterns to avoid peak traffic periods (a temporal shift; Slabbekoorn and Peet 2003; Brumm 2004; Warren et al. 2006; Wood and Yezerinac 2006; Fuller et al. 2007; Parris and Schneider 2009; Potvin et al. 2011). Recent studies have shown that songbirds can shift the frequency of their vocal signals as a highly plastic, short-term behavioural response to urban noise (Bermúdez-Cuamatzin et al. 2009; Halfwerk and Slabbekoorn 2009; Potvin and Mulder 2013). The southern brown tree frog *Litoria ewingii* from south-eastern Australia and the bow-winged grasshopper *Chorthippus biguttulus* from Germany have been observed to call at a higher pitch in traffic noise (Parris et al. 2009; Lampe et al. 2012), while whales have demonstrated frequency and amplitude shifts in their songs when exposed to high levels of noise from passing boats (Lesage et al. 1999; Scheifele et al. 2005; Parks et al. 2007). In the case of the bow-winged grasshopper, developmental plasticity rather than individual behavioural plasticity appears to be the mechanism for frequency shifts in animals from roadside habitats (Lampe et al. 2014; Figure 3.5).

Population-level impacts of traffic noise on animals are more difficult to observe. An early study in The Netherlands found lower reproductive rates and lower population densities of birds in habitats close to roads (Reijnen and Foppen 1994; Reijnen et al. 1995, 1996), although these results may have been confounded by habitat differences between the roadside study sites and control sites. Traffic noise was proposed as the primary cause of reduced breeding success in male willow warblers *Phylloscopus trochilus* with territories near busy roads; these birds had difficulty attracting and maintaining a mate (Reijnen and Foppen 1994). A more recent study of great tits *Parus major* revealed similar results; females laid fewer eggs in noisy areas and the number of fledglings was also reduced (Halfwerk et al. 2011). The probability of detecting the grey shrike-thrush *Colluricincla harmonica* and the grey fantail *Rhipidura fuliginosa* in roadside habitats declined with increasing traffic noise on the Mornington Peninsula, south-eastern Australia (Parris and Schneider 2009). Across 58 study sites, the predicted probability of detecting the grey shrike-thrush on a visit to a site ranged from 86% (95% CI: 61–100%) at the quietest site to 10% (0–28%) at the noisiest site. This relationship held when

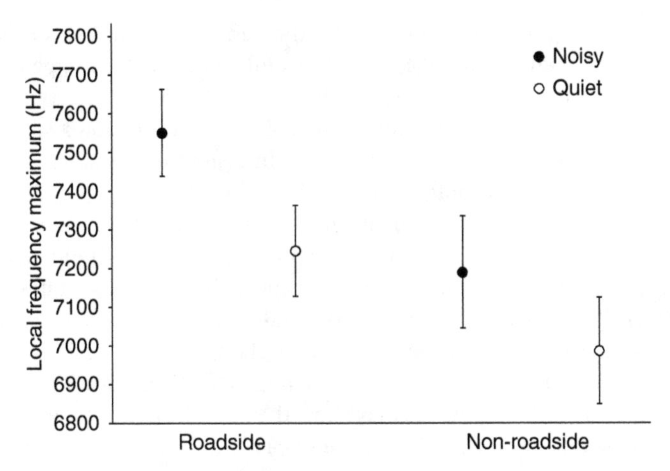

Figure 3.5 Frequency shift in the call of the bow-winged grasshopper *Chorthippus biguttulus* in noisy conditions (local frequency maxima in Hz), showing results for individuals from roadside and non-roadside habitats; predicted effect ± SE for a male with overall mean body mass of 95.15 mg. Lampe et al. 2014, Figure 2. Reproduced with permission from John Wiley & Sons.

differences in vegetation type and width, landscape context, and traffic volume (the number of passing vehicles) between sites were considered, suggesting that the occurrence of this species decreased in response to noise. An experiment with a "phantom road" (where traffic noise was introduced to a roadless landscape in Idaho, USA) found a lower abundance of birds across multiple species on days when noise was played (McClure et al., 2013), demonstrating that the birds were responding to the noise itself rather than any other feature of a road.

In the woodlands of New Mexico, USA, some species of birds nested less frequently at sites affected by the noise of natural-gas compressors than at quiet control sites, while other species nested more frequently (Francis et al. 2009). All nests of the black-headed grosbeak *Pheucticus melanocephalus* and 22 of 23 nests of the mourning dove *Zenaida macroura* were located at quiet sites. Both these species communicate using acoustic signals in the lower part of the frequency spectrum and would therefore suffer substantial acoustic interference from the low-frequency compressor noise (Francis et al. 2009). Interestingly, birds that nested at noisy sites experienced lower rates of nest predation and higher reproductive success than those at quiet sites; this relationship held both within species and across all species in the nesting community. The main nest predator in the study area – the western scrub-jay *Aphelocoma californica* – was much less likely to occur at noisy sites. This study shows that anthropogenic noise can affect not only the local abundance of certain species, but also the interactions between species, leading to an indirect, positive effect on some populations in noisy areas.

Artificial night lighting in cities changes the behaviour of animals in a variety of ways (Longcore and Rich 2004; Eisenbeis and Hänel 2009; Kempenaers et al. 2010). However, the majority of research into the biological effects of artificial night lighting has focused on individuals, with limited information available about population-level effects (Gaston et al. 2015). Increased illumination can extend diurnal activities, such as foraging and singing, into the night. Additional time to gather food may improve survival rates of animals that can take advantage of this night-light niche, such as certain reptiles (Powell and Henderson 2008), although foraging under lights may in turn expose them to higher rates of predation. Conversely, animals that prefer dark conditions for foraging or hunting are disadvantaged in well-lit urban environments (Longcore and Rich 2004). Some fast-flying bats are attracted to lights and the high density of insect prey that gathers there, while slow-flying species may show the opposite trend (Rydell and Racey 1995). The lesser horseshoe bat *Rhinolophus hipposideros* of Europe flies slowly and is vulnerable to predation by diurnal birds. A recent experiment found that the installation of high-pressure sodium lights reduced activity of this species along flyways, and delayed its nightly commuting activity (Stone et al. 2009).

Artificial lighting can disorient animals that require dark conditions to navigate, such as migrating birds and insects or nesting and hatching turtles (Witherington and Bjornal 1991; Longcore and Rich 2004; Gauthreaux and Belser 2006; New 2007; Bourgeois et al. 2009). Birds are particularly disoriented by artificial lighting on overcast nights when visual cues such as the moon or stars are unavailable, and by white and red lights which interfere with their magnetic compass (Poot et al. 2008). It is estimated that millions of migrating birds are impacted by artificial night lighting each year, and many of these do not survive – they may collide with the lit structures or become trapped in the lighted area and continue flying around to the point of exhaustion (Poot et al. 2008; Eisenbeis and Hänel 2009).

A study of leatherback turtles in Pongara National Park, Gabon found that turtle hatchlings emerging from nests on the beach deviated towards artificial lights instead of heading straight out to sea (Bourgeois et al. 2009). The effect was most pronounced along the section of beach with the greatest exposure to lights from the capital city Libreville, and with few landward shape cues, such as vegetation and logs. This deviation from a straight path would incur an energetic cost, and the extra time spent reaching the ocean would increase the risk of predation on the hatchlings (Bourgeois et al. 2009). Insects attracted to artificial lights may perish in large numbers, with implications for the persistence of populations and even entire species (Eisenbeis and Hänel 2009; van Langevelde et al. 2011). Polarized light pollution is a particular kind of light pollution in which light reflecting off dark, shiny, human-made surfaces (such as roads, buildings, and plastic sheeting) is polarized in a similar way to light reflecting off dark water bodies (Horváth et al. 2009). Many animals, including at least 300 species of aquatic insects, perceive polarized light and use it to orient towards suitable aquatic habitats; they can be falsely attracted to artificial polarizing surfaces, which act as a powerful ecological trap (Horváth et al. 2009; see Box 3.2).

Box 3.2 Population Sinks and Ecological Traps

In ecology, the source-sink model (Pulliam 1988) describes population dynamics in a multi-patch system with at least one patch of high-quality habitat where births outweigh deaths (population growth is positive) and one patch of low-quality habitat where deaths outweigh births (population growth is negative). Immigration of excess individuals from the former (source) population can sustain the latter (sink) population, theoretically indefinitely. Thus, a source population is a net exporter of individuals while a sink population is a net importer. The source-sink model has interesting applications in altered urban environments. For example, the presence of an apparently stable population of a given species in an urban area may suggest that the species in question has successfully adapted to habitat conditions in the city (i.e., it is an urban adapter or exploiter; see Section 3.5). However, closer examination reveals that the population has a high mortality rate and/or low breeding success, and is actually being sustained by immigration from a source population on the city fringe. A recent study of cat predation on birds in Dunedin, New Zealand presents just this scenario: the number of birds killed by cats was substantial for all six species studied, and greater than the estimated city-wide population size of the grey fantail *Rhipidura fuliginosa* and the introduced song thrush *Turdus philomelos* (van Heezik et al. 2010).

Ecological traps (Dwernychuk and Boag 1972) occur when rapid environmental change leads organisms to select poor-quality or sink habitats over high-quality habitats. Consider that individuals select a suitable habitat based on environmental cues. If either habitat quality or the distribution of cues changes such that a cue no longer reliably indicates the presence of a suitable habitat, individuals may choose sub-optimal habitats and suffer reduced survival or reproductive success (Schlaepfer et al. 2002). For example, the attraction of aquatic insects to artificial sources of polarizing light may lead them into an ecological trap, where they lay their eggs on shiny cement floors or black plastic sheeting rather than in water, or perish in pools of waste oil (reviewed by Horváth et al. 2009). An ecological trap focuses on habitat selection and is just one of a range of evolutionary traps (Schlaepfer et al. 2002). Organisms rely on environmental cues to make a variety of other decisions, such as when to reproduce, with whom to mate and when to migrate or emerge from hibernation. Environmental changes associated with urbanization may lead to a variety of evolutionary traps that could act in concert with other, more direct effects of urbanization, to the detriment of populations and species (Schlaepfer et al. 2002).

3.3 Biological introductions and invasions

3.3.1 Plants and fungi

Urban environments are characterized by the presence of many non-native species. While the majority of these species are deliberately or accidentally introduced by people, others have moved into cities and towns unaided to exploit a certain niche or resource. Cities around the world harbour many introduced plants, including garden favourites such as roses, camellias, daffodils, jacarandas, bougainvilleas, jasmine and frangipani, food plants such as tomatoes, pumpkins and culinary herbs, as well as many "weeds" of garden and agricultural origin. Introduced plants account for much of the observed pattern of high plant diversity in cities (Porter et al. 2001; Hope et al. 2003; Thompson et al. 2003; Zipperer and Guntenspergen 2009), while native plants originally present in an area may become locally or even globally extinct following urbanization (Thompson and Jones 1999; Williams et al. 2005; Kuhn and Klotz 2006; Hahs et al. 2009). Some exotic garden plants must be tended very carefully to survive outside their natural geographic and climatic range, while others thrive with little attention. Species in the latter group that are well suited to local conditions, able to colonize new areas following disturbance, have a high tolerance of disturbance, produce large numbers of seeds that are readily dispersed by wind or animals, and/or can reproduce vegetatively are most likely to make the transition from garden favourite to naturalized populations (Baker 1974; Buist et al. 2000; Lake and Leishman 2004). A variety of factors then influence whether a naturalized population of a species becomes invasive in a particular location, including residence time, climate, dispersal traits and propagule pressure (Richardson and Pyšek 2012).

Areas of remnant native vegetation, parks and gardens, industrial areas (both active and abandoned) and vacant blocks of land are all subject to invasion by exotic plants as well as native ruderal species. In a recent study of the alien flora of Europe, industrial habitats and managed parks and gardens supported the greatest number of introduced plant species of the ten habitat types surveyed (Lambdon et al. 2008). Invasion by exotic species in urban areas (as elsewhere) is facilitated by mechanical disturbance of the soil, the addition of nutrients such as phosphorus and nitrogen, and the fragmented nature of the human-dominated landscape (Riley and Banks 1996; Cilliers et al. 2008). Exotic fungi are also common in urban habitats, introduced with mulch or as symbionts of non-native plant species (Newbound 2008).

Urban streams, ponds and wetlands are often infested with non-native aquatic species that have been introduced deliberately from garden ponds or fish tanks, or washed into waterbodies from the upstream urban watershed (catchment) as seeds or vegetative material. Hussner (2009) found that four species of aquatic weeds in Europe have high relative growth rates that are further enhanced by high nutrient levels, such as those commonly encountered in urban waterways. All four species can regenerate vegetatively, and in the case of water primrose *Ludwigia grandiflora* and parrot feather *Myriophyllum aquaticum*, a new plant can regenerate from a single leaf (Hussner 2009). Depending on growth form, aquatic weeds can form dense, floating mats on the surface of urban streams, ponds and lakes, displacing native aquatic plants and reducing the oxygen content of the water (e.g., floating pennywort *Hydrocotyle ranunculoides*). Alternatively, aquatic weeds can develop dense stands, rooted in the substrate, that also exclude native species (e.g., *Typha* spp.; Ruiz-Avila and Klemm 1996; Zedler and Kercher 2004; Hussner 2009). Thus, aquatic weeds outcompete natives in urban waterways by growing faster in high-nutrient conditions, physically crowding or shading out other plants, and reproducing copiously.

Smooth cordgrass *Spartina alterniflora*, an invasive species from North America, has been deliberately introduced to islands in the Yangtze River delta near Shanghai, China, to enhance the accumulation of sediments on tidal mudflats and accelerate expansion of the islands (Chen et al. 2008). Population and economic growth in the Shanghai region has increased demand for land that can be used for agriculture and housing, while construction of large dam projects upstream has reduced the volume of sediment carried by the Yangtze River. However, the introduced *Spartina* is rapidly outcompeting native plants, including the sea bulrush *Scirpus mariqueter* and the common reed *Phragmites australis*. Wetlands dominated by *S. mariqueter* provide significant habitats for waterfowl and migratory neotropical birds, and are recognised by RAMSAR as wetlands of international importance (Chen et al. 2008). A study in the wetlands associated with the island of Jiuduansha found that all guilds of birds (perching birds, shallow-water foragers, dabbling ducks, moist-soil foragers and gulls) avoided areas dominated by *Spartina* (Ma et al. 2007). Despite being listed as a harmful invasive species by the State Environmental Protection Administration of China, *Spartina* is still being planted in the Yangtze River estuary (Ma et al. 2007). The demand for new land and the superior ability of communities dominated by *Spartina* to sequester carbon (compared to native vegetation communities) provide strong economic incentives to continue with this practice (Chen et al. 2008).

3.3.2 Animals

As well as a cornucopia of introduced plants and fungi, cities often support high densities of introduced animals. Mammalian predators such as cats, dogs and foxes are almost ubiquitous in urban areas; introduced geckoes cling to the walls of houses; exotic fish and turtles are found in many city ponds and streams; the offspring of escaped domestic parrots squawk in the trees of neighbourhood parks; exotic earthworms burrow in the soil of urban forests; while non-native marine animals such as polychaete worms, bryozoans, sea stars and mussels inhabit ports and harbours adjacent to cities (Steinberg et al. 1997; Hewitt et al. 2004; Beckerman et al. 2007). For example, 72 species of introduced animals have been found in Port Phillip Bay in southern Australia, along with 29 species of uncertain origin. Many of these have arrived on the hulls or in the ballast water of ships travelling from distant international ports (Hewitt et al. 2004). In urban environments, some introduced animals exist at low densities and without obvious effects on native species or ecosystem function. In contrast, others interact strongly with native species through competition for resources, direct predation, or more complex processes such as predator-mediated competition.

Many cities around the world host large populations of introduced birds, including blackbirds, starlings, sparrows and rock doves. The common myna *Acridotheres tristis*, native to Asia, has been introduced to Australia, New Zealand, South Africa, The Middle East, Europe and North America as well as many islands, including Hawaii and Madagascar (Peacock et al. 2007; BirdLife International 2012). It can occur at high densities in suburban areas and urban nature reserves, with population densities >100 birds/km^2 recorded in Canberra, Australia (Pell and Tidemann 1997a). Strongly territorial, it has been listed as one of the world's worst

100 invasive species by the IUCN (Lowe et al. 2000). Common mynas roost communally and nest in tree hollows (including artificial nest boxes), and compete aggressively with native species of birds for roost and nest sites (Pell and Tidemann 1997b). They displace breeding pairs of other birds, and have been known to physically eject nests and chicks from hollows (Pell and Tidemann 1997b; Blanvillain et al. 2003). They are therefore likely to reduce the breeding success and population size of native birds in urban areas where natural and/or artificial hollows are in limited supply (Pell and Tidemann 1997b; Blanvillain et al. 2003).

Introduced predators can reach high densities in both terrestrial and aquatic habitats in cities and may have a substantial impact on populations of the native species they prey upon. For example, a recent survey estimated that 9 million domestic cats in Britain captured and brought home 92 million prey items (95% CI: 85–100) over a five-month period, including 57 million mammals, 27 million birds and 5 million reptiles and amphibians (Woods et al. 2003). While these numbers seem high, they equate to only one prey item per cat every two weeks. The prey items recorded included 20 species of mammals and 44 species of birds, some of which are of conservation concern (e.g., the water shrew and yellow-necked mouse). Cats with bells attached to their collars and those kept indoors at night tended to catch fewer mammals, but the latter group caught more reptiles and amphibians while outdoors during the day (Woods et al. 2003; Figure 3.6). A more recent study in Bristol, UK attempted to quantify the impact of cat predation on populations of urban birds (Baker et al. 2008). Despite difficulties in accurately measuring the density and reproductive rate of birds and the number of individuals taken by cats, predation by domestic cats is likely to be having a significant impact on populations of common urban birds such as the dunnock *Prunella modularis*, the European robin *Erithacus rubecula* and the winter wren *Troglodytes troglodytes*. Across four study areas, the estimated number of individuals of these species killed exceeded the estimated number of juveniles fledged (Baker et al. 2008). In instances such as these, it is likely that urban bird populations are being maintained by immigration, and that urban areas are acting as population sinks (see Box 3.2).

The presence of introduced predators may also lead to more complex inter-specific interactions in urban environments. Modelling by Beckerman et al. (2007) indicated that the high density of cats in urban parts of Britain would be sufficient to reduce bird populations through low levels of predation combined with sub-lethal effects on fecundity. Even a small reduction in fecundity resulting from behavioural changes in the presence of cats could account for the dramatic declines in urban populations of common bird species observed in the last 30 years (Beckerman et al. 2007). The red fox *Vulpes vulpes* is another predator that has been widely introduced outside its native range and is commonly found in urban areas. This species was first introduced to California, USA in 1885, but its distribution remained restricted until a rapid expansion beginning in the 1970s (Lewis et al. 1999). Red foxes were common in Orange County in the 1990s,

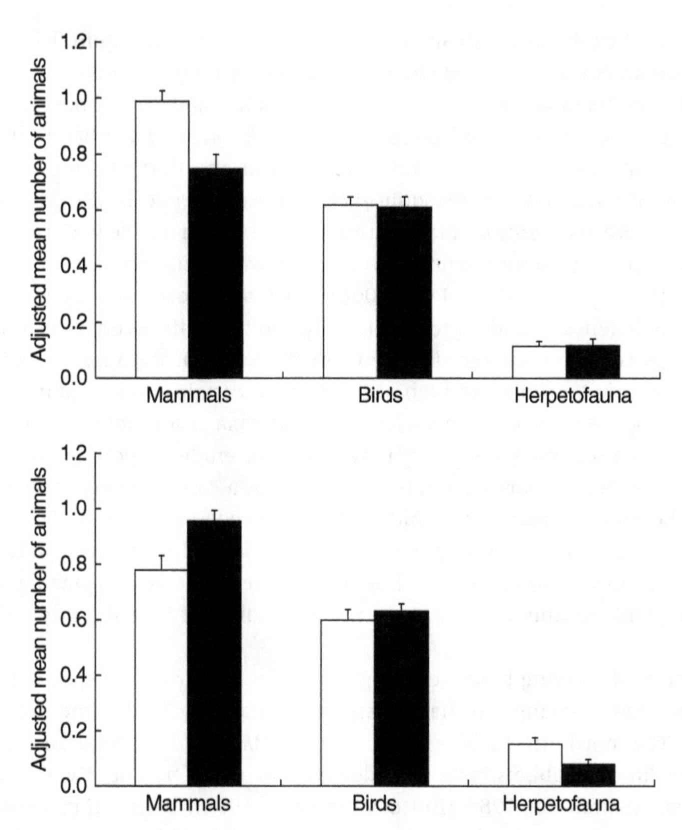

Figure 3.6 Prey items of 282 domestic cats in Great Britain: adjusted means (± S.D.) of $\log_{10}$-transformed numbers of mammals, birds and herpetofauna brought home by (a) cats that wore bells (black columns) and those that did not (white columns); (b) cats that were allowed out at night (black columns) and those that were not (white columns). Woods et al. 2003, Figures 5 and 6. Adapted with permission of John Wiley & Sons.

where urban development has reduced habitat suitable for native predators while creating open spaces such as parks, golf courses and cemeteries that provide refuge for foxes. Lewis et al. (1999) present this as an urban example of mesopredator release (*sensu* Soulé et al. 1988), where urbanization has reduced the habitat available for the top predators in this system (coyotes *Canis latrans* and mountain lions *Felis concolor*), allowing lower-tier mesopredators such as the introduced red fox to become unusually abundant (see Box 4.2 for further discussion of trophic relationships in urban environments). The red fox is a generalist predator, and is considered a threat to a range of smaller native fauna in California including the endangered light-footed clapper rail *Rallus longirostrus levipes* and the San Joaquin kit fox *Vulpes macrotis mutica* (Lewis et al. 1999).

Introduced freshwater fish such as carp, goldfish, and mosquitofish are common in urban ponds, lakes and streams, where they prey on native invertebrates, fish and amphibian larvae (Lowe et al. 2000; Hamer and McDonnell 2008). Two species of mosquitofish, *Gambusia affinis* and *G. holbrooki*, have been introduced worldwide from eastern North America, often in an effort to reduce mosquito larvae and the incidence of mosquito-borne diseases (Pyke 2008). Mosquitofish have a very broad environmental tolerance, occurring in a wide variety of aquatic habitats with still or slow-moving water, and at water temperatures ranging from 0–45 °C (Cherry et al. 1976; Pyke 2008). They are short-lived species that can occur at high densities and reproduce rapidly, but their effectiveness at controlling mosquitoes is uncertain (see discussion in Pyke 2008). Because of their intermediate trophic position and high abundance in certain urban aquatic habitats, *Gambusia* spp. can impact on a variety of native taxa at a number of trophic levels. Through direct predation, they may reduce or eradicate populations of other small fish, amphibian larvae, and invertebrates; as a consequence, competitors or prey of these species may benefit indirectly (reviewed in Pyke 2008). For example, Jassby et al. (1977a,1977b) observed selective predation by *Gambusia* on large zooplankton during a laboratory experiment, which led to decreased grazing pressure on phytoplankton and an increase in the abundance of bacteria, phytoplankton and rotifers.

As well as disrupting food webs, high densities of introduced species in urban areas can lead to changes in habitat structure that may be detrimental to other species. The northern Pacific seastar *Asterias amurensis*, a generalist predator in soft-sediment habitats, was first detected in the Derwent River estuary in Tasmania, Australia in 1986 (Buttermore et al. 1994). It has since become the dominant invertebrate predator in this system with densities ≤46 individuals/m^2, and is considered a major threat to benthic marine communities and commercial shellfisheries (Ross et al. 2002). A manipulative experiment demonstrated the substantial predation pressure that *A. amurensis* places on native bivalves. After 10 weeks, the mean density of recently-settled juveniles of the thin-ribbed cockle *Fulvia tenuicostata* was reduced from around 300/m^2 to 35/m^2 in the presence of seastars at natural densities, and to 17/m^2 in cage inclusions with a density of 1 seastar/m^2 (Ross et al. 2002). Conversely, the mean size of *F. tenuicostata* increased substantially where seastars were excluded. *Asterias amurensis* has also been implicated in the decline of the endangered spotted handfish *Brachionichthys hirsutus* in Tasmania, either through direct predation on its egg masses, or predation on the stalked ascidians that provide its principal spawning substrate (Bruce and Green 1998). *Brachionichthys hirsutus* is a small, slow-moving, benthic fish that 'walks' on its pectoral and pelvic fins which resemble human hands (Figure 3.7). It has a very restricted distribution and its decline during the 1980s coincided with the establishment and expansion of seastar populations in the Derwent River estuary (Bruce and Green 1998).

Figure 3.7 The endangered spotted handfish *Brachionichthys hirsutus* is endemic to the Derwent River Estuary, Tasmania, Australia. This species is threatened by the introduced northern Pacific seastar *Asterias amurensis*. CSIRO. https://commons.wikimedia.org/wiki/File%3ACSIRO_ScienceImage_10_The_Endangered_Spotted_Handfish.jpg. Used under CC BY 3.0 http://creativecommons.org/licenses/by/3.0.

3.4 Human disturbance

Urban-dwelling humans disturb other urban-dwelling organisms in many ways: they collect fungi, plants and animals for food or to sell in markets; move them from one place to another; dig up plants they do not like; mow and/or trample the ones they do; collect rocks and fallen timber that provide habitat for invertebrates and vertebrates; and alter the behaviour of a wide range of animals just with their presence. In general, the intensity of human disturbance in urban areas increases with human population density, but it also varies with cultural practices, social attitudes and the socio-economic status of a neighbourhood. Edible fungi, plants and animals, as well as standing and fallen timber that can be used for fuel, are likely to be under much greater pressure from humans in an impoverished neighbourhood than in an affluent one (e.g., Stoian 2005).

Areas of remnant vegetation in cities are often subject to trampling by humans, with a range of consequences including compaction of the soil, reduced root and stem growth, reduced plant cover and a decline in the species richness of the ground layer or understorey (reviewed by Florgård 2009). The impacts of trampling on understorey vegetation vary with soil type, soil depth and moisture levels, as well as the composition and structure of the vegetation community (Florgård 2009; Hill and Pickering 2009). Experimental trampling trials in a sub-tropical urban reserve on the Gold Coast, Australia found that vegetation with a fern understorey had low resistance to trampling, with the relative height and cover of the understory reduced following as few as ten passes (Hill and Pickering 2009). Understorey vegetation dominated by tussock grasses and introduced lawn grasses had moderate and high resistance to trampling, respectively. Physical damage to plants and soil caused by human trampling can stunt plant growth, kill individual plants, and reduce the likelihood of successful recruitment of juvenile plants. Thus, trampling can impact on vital rates to the extent that populations of certain plants are excluded from trampled sites (Figure 3.1). Conversely, plant species that are tolerant of this type of disturbance may flourish in heavily-trampled areas (Hamberg et al. 2008).

Trampling of plants and animals in the intertidal zone is common in coastal areas with regular human foot traffic, with differential impacts on the cover or density of different species (Keough and Quinn 1998). Over time, this could lead to the local extinction of species susceptible to repeated trampling. Another human disturbance – collection of sandstone rocks for landscaping urban gardens – is implicated in the decline of the broad-headed snake *Hoplocephalus bungaroides* in south-eastern Australia (Webb and Shine 2000). Removal of these rocks reduces the availability of diurnal retreat sites for the snake and its important prey species, the velvet gecko *Oedura lesueurii*. An experiment in Morton National Park in southern New South Wales demonstrated that concrete pavers could provide suitable artificial retreat sites for these reptiles at degraded sites (Webb and Shine 2000). Collection of bush rock from public lands in News South Wales is now illegal, and bush-rock removal is listed as a key threatening process under the New South Wales *Threatened Species Conservation Act 1995*.

A more widespread, but less obviously detrimental, activity of humans in urban areas is walking, both with and without domestic dogs. There has been considerable debate about the potential impact of human walkers and their dogs on native wildlife, particularly birds. Disturbance by human pedestrians (without dogs) in urban parks in Madrid was found to decrease the diversity and abundance of birds, and the probability that individual species would be present in a given park (Fernández-Juricic 2000). An experimental manipulation at 90 woodland sites on the urban fringe of Sydney, Australia investigated the impact on birds of one person walking a dog and one person walking alone, compared to control sites without walking (Banks and Bryant 2007). The sites with dog walking had 35% fewer bird species and 41% fewer individual birds than control sites in the ten minutes immediately after the human and dog walked past (Banks and Bryant 2007). Sites with a human walking alone also had lower bird diversity and abundance than control sites, but the effect was smaller than at dog-walking sites. Ground-dwelling birds appeared to be most affected by dog walking, with half the species recorded in control sites absent from the dog-walking sites. Interestingly, the effect of dogs occurred even in areas where dog walking is common, suggesting that the local birds do not become habituated to continued disturbance (Banks and Bryant 2007). Bike riding on trails through urban forests may also disturb birds and other wildlife, leading to reduced population densities (e.g., George and Crooks 2006; Thompson 2015).

A recent study in the UK found important effects of human walkers and their dogs on the Dartford warbler *Sylvia undata* (Murison et al. 2007). This bird has a limited distribution in the UK, where it breeds exclusively in the coastal heathlands of southern England. Many coastal heathlands in England are near urban centres, and are frequented by people who wish to walk their dogs, watch birds and appreciate nature (Murison et al. 2007; Underhill-Day and Liley 2007). The study demonstrated that disturbance by human visitors and their dogs delayed breeding in *S. undata* by up to six weeks, leading to a reduced number of broods per year and fewer chicks fledged per breeding pair (Murison et al. 2007). Two mechanisms were suggested to account for this pattern: disturbance delayed hatching dates, such that the growth period of chicks did not coincide with the period of optimal invertebrate prey density, thereby decreasing chick survival; and disturbance events directly interrupted foraging and chick-feeding by adults (Murison et al. 2007).

In an example of human disturbance in marine habitats, the presence of boats (both commercial fishing boats and whale-watching boats) was found to change the behaviour of killer whales *Orcinus orca* in the Johnstone Strait, British Columbia, Canada (Williams et al. 2006). The whales spent less time feeding and rubbing their bodies on smooth pebble beaches, with the lost feeding opportunities estimated to reduce their energy intake by 18%. Heavy boat traffic in marine habitats close to cities could impact on a range of cetaceans (e.g., Dans et al. 2008; Stockin et al. 2008), potentially leading to population declines where disturbance is frequent enough to substantially interfere with their feeding activities.

3.5 Stochastic effects on populations in urban environments

As mentioned in Chapter 2, three types of stochasticity are known to affect the birth and death rates of individuals within a population; environmental stochasticity, demographic stochasticity and catastrophes (Lande 1993). Small, isolated populations of native species in fragments of habitat within or adjacent to cities are vulnerable to environmental and demographic stochasticity (Bolger et al. 1997; Kéry et al. 2003; Sattler et al. 2010). Random fluctuations in environmental conditions that decrease birth rates and/or increase mortality may be sufficient to drive such populations to extinction. Similarly, a chance failure of reproduction over one or more breeding seasons and/or lower than average adult survival may result in the local extinction of small populations, with isolation also reducing the probability of successful recolonization in the future. Both small and large populations can be impacted by environmental catastrophes, such as wildfire, floods, disease, intense storms and extended periods of drought. While the risk of a particular catastrophe, such as wildfire or flood, may be reduced in urban centres through the management of fuel loads, active fire suppression, and structural modifications to waterways, it cannot be excluded entirely (e.g., Keeley et al. 1999; Nyambod 2010; Buxton et al. 2011). One benefit of habitat fragmentation in urban areas is a decreased probability that contagious processes such as fire and disease will spread between isolated populations, which are therefore less likely to suffer simultaneous extinctions in the event of a catastrophe (Fahrig 2003). Generally speaking, deterministic and stochastic processes work in concert to influence the growth rate of populations in urban environments and their probability of persistence or extinction over time.

3.6 Summary

Populations (and the species to which they belong) respond to urbanization in a diversity of ways. The processes of urbanization can be classified into three broad groups; biophysical processes, biological introductions and human disturbance. These processes alter the quantity and quality of resources that organisms depend on for survival, as well as their distribution in time and space. They can also change the interactions within and between species in both subtle and dramatic ways. These changes impact on the four vital rates – births, deaths, immigration and emigration – leading to population growth or decline, which in turn may further alter intra- and inter-specific interactions within the system. Thus, the processes of urbanization may ultimately result in the local extinction of populations and/or species that were present prior to the construction of a city, reduction in the abundance of others, increased abundance of certain native and non-native species that can thrive in urban conditions, and changes in the global distribution of species. In the following chapter, I examine the varied responses of ecological communities to urbanization.

Study questions

1 Name the four processes or vital rates that underlie the dynamics of populations.
2 By what mechanisms and pathways are these population-level processes affected by the biophysical processes of urbanization?
3 What is a metapopulation? How do metapopulations operate in cities?
4 Urban noise and light regimes have important effects on native species – discuss.
5 What is an ecological trap? How would you recognize one?
6 How do human disturbance and biological introductions impact on populations of native species in urban environments?

References

Agneta, M. and Burton, S. (1990) Terrestrial and aquatic bryophytes as monitors of environmental contaminants in urban and industrial habitats. *Botanical Journal of the Linnean Society*, **104**, 267–280.

Amorim, M.C.P. and Vasconcelos, R.O. (2008) Variability in the mating calls of the Lusitanian toadfish *Halobatrachus didactylus*: cues for potential individual recognition. *Journal of Fish Biology*, **73**, 1267–1283.

ANZECC (2000) *Australian and New Zealand Guidelines for Fresh and Marine Water Quality Volume 1, The Guidelines*. Australian and New Zealand Environment and Conservation Council and the Agriculture and Resource Management Council of Australia and New Zealand, Australian Government, Canberra.

Asfaw, Z. and Tadesse, M. (2001) Prospects for sustainable use and development of wild food plants in Ethiopia. *Economic Botany*, **55**, 47–62.

Baker, H.G. (1974) The evolution of weeds. *Annual Review of Ecology and Systematics*, **5**, 1–24.

Baker, L.A., Brazel, A.J., Selover, N. *et al.* (2002) Urbanization and warming of Phoenix (Arizona, USA): Impacts, feedbacks and mitigation. *Urban Ecosystems*, **6**, 183–203.

Baker, P.J., Molony, S.E., Stone, E. *et al.* (2008) Cats about town: is predation by free-ranging pet cats *Felis catus* likely to affect urban bird populations? *Ibis*, **150**, 86–99.

Banks, P.B. and Bryant, J.V. (2007) Four-legged friend or foe? Dog walking displaces native birds from natural areas. *Biology Letters*, **3**, 611–613.

Banks, W.A., Altmann, J., Sapolsky, R.M. *et al.* (2003) Serum leptin as a marker for a syndrome X-like condition in wild baboons. *The Journal of Clinical Endocrinology and Metabolism*, **88**, 1234–1240.

Barber, J.R., Crooks, K.R., and Fristrup, K.M. (2010) The costs of chronic noise exposure for terrestrial organisms. *Trends in Ecology and Evolution*, **25**, 180–189.

Batary, P. and Baldi, A. (2004) Evidence of an edge effect on avian nest success. *Conservation Biology*, **18**, 389–400.

Bates, A.J., Sadler, J.P., Fairbrass, A.J. *et al.* (2011) Changing bee and hoverfly pollinator assemblages along as urban-rural gradient. *PLoS ONE* **6**, e23459. doi:10.1371/journal.pone.0023459

Baur, B. and Baur, A. (1993) Climatic warming due to thermal-radiation from an urban area as possible cause for the local extinction of a land snail. *Journal of Applied Ecology*, **30**, 333–340.

Beale, C.M. and Monaghan, P. (2004) Human disturbance: people as predation-free predators? *Journal of Applied Ecology*, **41**, 335–343.

Beckerman, A.P., Boots, M., and Gaston, K.J. (2007) Urban bird declines and the fear of cats. *Animal Conservation*, **10**, 320–325.

Beckmann, J.P. and Berger, J. (2003) Rapid ecological and behavioural changes in carnivores: the response of black bears (*Ursus americanus*) to altered food. *Journal of Zoology*, **261**, 207–212.

Bee, M.A. and Swanson, E.M. (2007) Auditory masking of anuran advertisement calls by road traffic noise. *Animal Behaviour*, **74**, 1765–1776.

Beebee, T.J. (1979) Habitats of the British amphibians (2): suburban parks and gardens. *Biological Conservation*, **15**, 241–257.

Bermúdez-Cuamatzin, E., Rios-Chelen, A.A., Gil, D. *et al.* (2009) Strategies of song adaptation to urban noise in the house finch: syllable pitch plasticity of differential syllable use? *Behaviour*, **146**, 1269–1286.

Beyene, A., Legesse, W., Triest, L. *et al.* (2009) Urban impact on ecological integrity of nearby rivers in developing countries: the Borkena River in highland Ethiopia. *Environmental Monitoring and Assessment*, **153**, 461–476.

Bhattacharya, M., Primack, R.B., and Gerwein, J. (2003) Are roads and railroads barriers to bumblebee movement in a temperate suburban conservation area? *Biological Conservation*, **109**, 37–45.

BirdLife International. (2012) *Acridotheres tristis*. The IUCN Red List of Threatened Species. Version 2015.2. www.iucnredlist.org (accessed on 20/07/2015)

Blanvillain, C., Salducci, J.M., Tutururai, G. *et al.* (2003) Impact of introduced birds on the recovery of the Tahiti Flycatcher (*Pomarea nigra*), a critically endangered forest bird of Tahiti. *Biological Conservation*, **109**, 197–205.

Blickley, J.L., Blackwood, D.L., and Patricelli, G.L. (2012a) Experimental evidence for the effects of chronic anthropogenic noise on abundance of greater sage-grouse at leks. *Conservation Biology*, **26**, 461–471.

Blickley, J.L., K. Word, Krakauer, A.H. *et al.* (2012b) The effect of experimental exposure to chronic noise on fecal corticosteroid metabolites in lekking male greater sage-grouse (*Centrocercus urophasianus*). *PLoS ONE* **7**(11): e50462. doi: 10.1371/journal.pone.0050462

Bolger, D.T., Alberts, A.C., Sauvajot, R.M. *et al.* (1997) Response of rodents to habitat fragmentation in coastal southern California. *Ecological Applications*, **7**, 552–563.

Booker, D.J. (2003) Hydraulic modelling of fish habitat in urban rivers during high flows. *Hydrological Processes*, **17**, 577–599.

Both, C., van Asch, M., Bijlsma, R.G. *et al.* (2008) Climate change and unequal phenological changes across four trophic levels: constraints or adaptations? *Journal of Animal Ecology*, **78**, 73–83.

Bourgeois, S., Gilot-Fromont, E., Viallefont, A. *et al.* (2009) Influence of artificial lights, logs, and erosion on leatherback sea turtle hatchling orientation at Pongara National Park, Gabon. *Biological Conversation*, **142**, 85–93.

Brattstrom, B.H. and Bondello, M.C. (1983) Effects of off-road vehicle noise on desert vertebrates, in *Environmental Effects of Off-road Vehicles* (eds R.H. Webb and H.H. Wilshire), Springer-Verlag, New York, pp. 167–206.

Bruce B.A. and Green, M.A. (1998) *The Spotted Handfish 1999–2001 Recovery Plan*. Australian Government, Canberra. http://www.environment.gov.au/resource/spotted-handfish-1999-2001-recovery-plan

Brumm, H. (2004) The impact of environmental noise on song amplitude in a territorial bird. *Journal of Animal Ecology*, **73**, 434–440.

Buist, M., Yates, C.J., and Ladd, P.G. (2000) Ecological characteristics of *Brachychiton populneus* (Sterculiaceae) (kurrajong) in relation to the invasion of urban bushland in south-western Australia. *Austral Ecology*, **25**, 487–496.

Bunnell, F.L. (1999) What habitat is an island?, in *Forest Fragmentation: Wildlife and Management Implications* (eds J.A. Rochelle, L.A. Lehmann, and J. Wisniewski), Brill, Leiden, pp. 1–31.

Burgess, J., Harrison, C.M., and Limb, M. (1988) People, parks, and the urban green: a study of popular meanings and values for open spaces in the city. *Urban Studies*, **25**, 455–473.

Buttermore, R.E., Turner, E., and Morrice, M.G. (1994) The introduced northern Pacific seastar *Asterias amurensis* in Tasmania. *Memoirs of the Queensland Museum*, **36**, 21–25.

Buxton, M., Haynes, R., Mercer, D. *et al.* (2011) Vulnerability to bushfire risk at Melbourne's urban fringe: the failure of regulatory land use planning. *Geographical Research*, **49**, 1–12.

Chamberlain, D.E., Cannon, A.R., Toms, M.P. *et al.* (2009) Avian productivity in urban landscapes: a review and meta-analysis. *Ibis*, **151**, 1–18.

Chen, J., Zhao, B., Ren, W. *et al.* (2008) Invasive *Spartina* and reduced sediments: Shanghai's dangerous silver bullet. *Journal of Plant Ecology*, **1**, 79–84.

Cherry, D.S., Rodgers, J.H. Jr., Cairnes, J. Jr. *et al.* (1976) Responses of mosquitofish (*Gambusia affinis*) to ash effluent and thermal stress. *Transactions of the American Fisheries Society*, **105**, 686–694.

Cilliers, S.S., Williams, N.S.G., and Barnard, F.J. (2008) Patterns of exotic plant invasions in fragmented urban and rural grasslands across continents. *Landscape Ecology*, **23**, 1243–1256.

Clark, T.N., Lloyd, R., Wong, K.K. *et al.* (2002) Amenities drive urban growth. *Journal of Urban Affairs*, **24**, 493–515.

Clarke, G.P., White, P.C.L., and Harris, S. (1998) Effects of roads on badger *Meles meles* populations in south-west England. *Biological Conservation*, **86**, 117–124.

Close, D.C., Messina, G., Krauss, S.L. *et al.* (2006) Conservation biology of the rare species *Conospermum undulatum* and *Macarthuria keigheryi* in an urban bushland remnant. *Australian Journal of Botany*, **54**, 583–593.

Contesse, P., Hegglin, D., Gloor, S. *et al.* (2004) The diet of urban foxes (*Vulpes vulpes*) and the availability of anthropogenic food in the city of Zurich, Switzerland. *Mammalian Biology*, **69**, 81–95.

Czech, B. and Krausman, P.R. (1997) Distribution and causation of species endangerment in the United States. *Science*, **277**, 1116.

Czech, B., Krausman, P.R., and Devers, P.K. (2000) Economic associations among causes of species endangerment in the United States. *BioScience*, **50**, 593–601.

Dans, S.L., Crespo, E.A., Pedraza, S.N. *et al.* (2008) Dusky dolphin and tourist interaction: effect on diurnal feeding behaviour. *Marine Ecology Progress Series*, **369**, 287–296.

De la Puente, D., Ochoa, C., and Viejo, J.L. (2008) Butterflies killed on roads (Lepidoptera, Papilionoidea) in "El Regajal-Mar de Ontigola" Nature Reserve (Aranjuez, Spain). *XVII Bienal de la Real Sociedad Española de Historia Natural*, **17**, 137–152.

Dos Santos, M.E., Modesto, T., Matos, R.J. *et al.* (2000) Sound production by the Lusitanian toad fish, Halobatrachus didactylus. *The International Journal of Animal Sound and its Recording*, **10**, 309–321.

Drinnan, I.N. (2005) The search for fragmentation thresholds in a southern Sydney suburb. *Biological Conversation*, **124**, 339–349.

Durant, J.M., Hjermann, D.Ø., Ottersen, G. *et al.* (2007) Climate and the match or mismatch between predator requirements and resource availability. *Climate Research*, **33**, 271–283.

Dwernychuk, L.W. and Boag, D.A. (1972) Ducks nesting in association with gulls – an ecological trap? *Canadian Journal of Zoology*, **50**, 559–563.

Eisenbeis, G. and Hänel, A. (2009) Light pollution and the impact of artificial night lighting on insects, in *Ecology of Cities and Towns* (eds M.J. McDonnell, A.K. Hahs, and J.H. Breuste), Cambridge University Press, Cambridge, pp. 243–263.

Erbe, C. (2002) Underwater noise of whale-watching boats and potential effects on killer whales (*Orcinus orca*), based on an acoustic impact model. *Marine Mammal Science*, **18**, 394–418.

Fahrig, L. (2003) Effects of habitat fragmentation on biodiversity. *Annual Review of Ecology, Evolution and Systematics*, **34**, 487–515.

Faulkner, S. (2004) Urbanization impacts on the structure and function of forest wetlands. *Urban Ecosystems*, **7**, 89–106.

Fernández-Juricic, E. (2000) Avifaunal use of wooded streets in an urban landscape. *Conservation Biology*, **14**, 513–521.

Fischer, J. and Lindenmayer, D.B. (2006) Beyond fragmentation: the continuum model for fauna research and conservation in human-modified landscapes. *Oikos*, **112**, 473–480.

Florgård, C. (2009) Preservation of original natural vegetation in urban areas: an overview, in *Ecology of Cities and Towns: A Comparative Approach* (eds M.J. McDonnell, A.K. Hahs, and J.H. Breuste), Cambridge University Press, Cambridge, pp. 380–398.

Foster, E., Curtis, L.R., and Gundersen, D. (2014) Toxic Contaminants in the urban aquatic environment, in *Wild Salmonids in the Urbanizing Pacific Northwest* (eds J.A. Yeakley, K.G. Maas-Hebner, and R.M. Hughes), Springer, New York, pp. 123–144.

Fox, C.H. (2006) Coyotes and humans: can we coexist? *Proceedings of the Vertebrate Pest Conference*, **22**, 287–293.

Francis, C.D., Ortega, C.P., and Cruz, A. (2009) Noise pollution changes avian communities and species interactions. *Current Biology*, **19**, 1–5.

Frankham, R., Ballou, J.D., and Briscoe, D.A. (2002) *Introduction to Conservation Genetics*, Cambridge University Press, Cambridge.

Frankham, R., Ballou, J.D., and Briscoe, D.A. (2004) *A Primer of Conservation Genetics*, Cambridge University Press, Cambridge.

Fuller, R.A., Irvine, K.N., Devine-Wright, P. *et al.* (2007) Psychological benefits of greenspace increase with biodiversity. *Biology Letters*, **3**, 390–394.

Gadsdon, S.R., Dagley, J.R., Wolseley, P.A. *et al.* (2010) Relationships between lichen community composition and concentrations of NO_2 and NH_3. *Environmental Pollution*, **158**, 2553–2560.

Garland, T. (1984) Physiological correlates of locomotory performance in a lizard: an allometric approach. *American Journal of Physiology*, **247**, 806–815.

Gaston, K.J., Visser, M.E., and Hölker, F. (2015) The biological impacts of artificial light at night: the research challenge. *Philosophical Transactions of the Royal Society B* **370**, 20140133.

Gates, E. and Gysel, L.W. (1978) Avian nest dispersion and fledging success in field-forest ecotones. *Ecology*, **59**, 871–883.

Gauthreaux, S.A. and Belser, C.G. (2006) Effects of artificial night lighting on migrating birds, in *Ecological Consequences of Artificial Night Lighting* (eds C. Rich and T. Longcore), Island Press, Washington DC, pp. 67–93.

Gehrt, S.D. (2004) Ecology and management of striped skunks, raccoons, and coyotes in urban landscapes, in *Predators and People: From Conflict to Conservation* (eds N. Fascione, A. Delach, and M. Smith), Island Press, Washington DC, pp. 81–104.

George, S.L. and Crooks, K.R. (2006) Recreations and large mammal activity in an urban nature reserve. *Biological Conservation*, **133**, 107–117.

Gibbs, J.P. and Shriver, W.G. (2002) Estimating the effects of road mortality on turtle populations. *Conservation Biology*, **16**, 1647–1652.

Gilbert, O.L. (1969) The effect of SO_2 on lichens and bryophytes around Newcastle upon Tyne, in *Air Pollution: Proceedings of the First European Congress on the Influence of Air Pollution on Plants and Animals*, Center for Agricultural Publishing and Documentation, Wageningen, pp. 223–235.

Gilbert, O.L. (1971) Some indirect effects of air pollution on bark-living invertebrates. *Journal of Applied Ecology*, **8**, 77–84.

Giordani, P. (2007) Is the diversity of epiphytic lichens a reliable indicator of air pollution? A case study from Italy. *Environmental Pollution*, **146**, 317–323.

Grimm, N.B., Foster, D., Groffman, P. *et al.* (2008) The changing landscape: ecosystem responses to urbanization and pollution across climate and societal gradients. *Frontiers in Ecology and the Environment*, **6**, 264–272.

Hahs, A.K., McDonnell, M.J., McMarthy, M.A. *et al.* (2009) A global synthesis of plant extinction rates in urban areas. *Ecology Letters*, **12**, 1165–1173.

Hale, J.M., Heard, G.W., Smith, K.L. *et al.* (2013) Structure and fragmentation of growling grass frog metapopulations. *Conservation Genetics*, **14**, 313–322.

Halfwerk, W. and Slabbekoorn, H. (2009) A behavioural mechanism explaining noise-dependent frequency use in urban birdsong. *Animal Behaviour*, **78**, 1301–1307.

Halfwerk, W., Holleman, L.J.M., Lessells, C.M. *et al.* (2011) Negative impact of traffic noise on avian reproductive success. *Journal of Applied Ecology*, **48**, 210–219.

Hamberg, L., Lehvävirta, S., Malmivaara-Lämsä, M. *et al.* (2008) The effects of habitat edges and trampling on understorey vegetation in urban forests in Helsinki, Finland. *Applied Vegetation Science*, **11**, 83–86.

Hamer, A.J. and McDonnell, M.J. (2008) Amphibian ecology and conservation in the urbanising world: a review. *Biological Conservation*, **141**, 2432–2449.

Hanski, I.A. (1994) A practical model of metapopulation dynamics. *Journal of Animal Ecology*, **63**, 151–162.

Hanski, I.A. (1998) Metapopulation dynamics. *Nature*, **396**, 41–49.

Hanski, I.A. and Ovaskainen, O. (2002) Extinction debt at extinction threshold. *Conservation Biology*, **16**, 666–673.

Harper, M.J. (2005) Home range and den use of common brushtail possums (*Trichosurus vulpecula*) in urban forest remnants. *Wildlife Research*, **32**, 681–687.

Harper, M.J., McCarthy, M.A., and van der Ree, R. (2008) Resources at the landscape scale influence possum abundance. *Austral Ecology*, **33**, 243–252.

Hawksworth, D.L. (1970) Lichens as litmus for air pollution: a historical review. *International Journal of Environmental Studies*, **1**, 281–296.

Heard, G.W., McCarthy, M.A., Scroggie, M.P., Baumgartner, J.B., and Parris, K.M. (2013) A Bayesian model of metapopulation viability, with application to an endangered amphibian. *Diversity and Distributions*, **19**, 555–566.

Hedrick, P.W. and Kalinowski, S.T. (2000) Inbreeding depression in conservation biology. *Annual Review of Ecology and Systematics*, **31**, 139.

Hels, T. and Buchwald, E. (2001) The effect of road kills on amphibian populations. *Biological Conservation*, **99**, 331–340.

Hewitt, C.L., Campbell, M.L., Thresher, R.E. *et al.* (2004) Introduced and cryptogenic species in Port Phillip Bay, Victoria, Australia. *Marine Biology*, **144**, 183–202.

Hill, R. and Pickering, C. (2009) Difference in resistance of three subtropical vegetation types to experimental trampling. *Journal of Environmental Management*, **90**, 1305–1312.

Hitchings, S.P. and Beebee, T.J.C. (1998) Loss of genetic diversity and fitness in Common Toad (*Bufo bufo*) populations isolated by inimical habitat. *Journal of Evolutionary Biology*, **11**, 269–283.

Hope, D., Gries, C., Zhu, W. *et al.* (2003) Socioeconomics drive urban plant diversity. *Proceedings of the National Academy of Sciences of the United States of America*, **100**, 8788–8792.

Horváth, G., Kriska, G., Malik, P. *et al.* (2009) Polarized light pollution: a new kind of ecological photopollution. *Frontiers in Ecology and the Environment*, **7**, 317–325.

How, R.A. and Dell, J. (2000) Ground vertebrate fauna of Perth's vegetation remnants: impacts of 170 years of urbanization. *Pacific Conservation Biology*, **6**, 198–217.

Hussner, A. (2009) Growth and photosynthesis of four invasive aquatic plant species in Europe. *Weed Research*, **49**, 506–515.

Jassby, A., Dudzik, M., Rees, J. *et al.* (1977a) *Production cycles in aquatic microcosms*. U.S. Environmental Protection Agency Report EPA-600/7–77–077. Environmental Protection Agency, Washington DC.

Jassby, A., Rees, J., Dudzik, M. *et al.* (1977b) *Trophic structure modifications by planktivorous fish in aquatic microcosms*. U.S. Environmental Protection Agency Report EPA-600/7–77–096. Environmental Protection Agency, Washington DC.

Johnson, P.T.J., Hoverman, J.T., McKenzie, V.J. *et al.* (2013) Urbanization and wetland communities: applying metacommunity theory to understanding the local and landscape effects. *Journal of Applied Ecology*, **50**, 34–42.

Keeley, J.E., Fotheringham, C.J., and Morais, M. (1999) Re-examining fire suppression impacts on brushland fire regimes. *Science*, **284**, 1829–1832.

Keller, L.F. and Waller, D.M. (2002) Inbreeding effects in wild populations. *Trends in Ecology and Evolution*, **17**, 230–241.

Kempenaers, B., Borgstrom, P., Loes, P. *et al.* (2010) Artificial night lighting affects dawn song, extra-pair siring success, and lay date in songbirds. *Current Biology*, **20**, 1735–1739.

Kentula, M.E., Gwin, S.E., and Pierson, S.M. (2004) Tracking changes in wetlands with urbanization: sixteen years of experience in Portland, Oregon, USA. *Wetlands*, **24**, 734–743.

Keough, M.J. and Quinn, G.P. (1998) Effects of periodic disturbances from trampling on rocky intertidal algal beds. *Ecological Applications*, **8**, 141–161.

Kéry, M., Matthies, D., and Schmid, B. (2003) Demographic stochasticity in population fragments of the declining distylous perennial *Primula veris* (Primulaceae). *Basic and Applied Ecology*, **4**, 197–206.

Kight, C.R. and Swaddle, J.P. (2011) How and why environmental noise impacts animals: an integrative, mechanistic review. *Ecology Letters*, **14**, 1052–1061.

King, S.A. and Buckney, R.T. (2000) Urbanization and exotic plants in northern Sydney streams. *Austral Ecology*, **25**, 455–461.

Konrad, C.P. and Booth, D.B. (2005) Hydrologic changes in urban streams and their ecological significance. *American Fisheries Society Symposium*, **47**, 157–177.

Kuhn, I. and Klotz, S. (2006) Urbanization and homogenization – comparing the floras of urban and rural areas in Germany. *Biological Conservation*, **127**, 292–300.

Kuussaari, M., Bommarco, R., Heikkinen, R.K. *et al.* (2009) Extinction debt: a challenge for biodiversity conservation. *Trends in Ecology and Evolution*, **24**, 564–571.

Lake, J.C. and Leishman, M.R. (2004) Invasion success of exotic plants in natural ecosystems: the role of disturbance, plant attributes, and freedom from herbivores. *Biological Conservation*, **117**, 215–226.

Lambdon, P.W., Pysek, P., Basnou, C. *et al.* (2008) Alien flora of Europe: species diversity, temporal trends, geographical patterns, and research needs. *Preslia*, **80**, 101–149.

Lampe, U., Reinhold, K., and Schmoll, T. (2014) How grasshoppers respond to road noise: developmental plasticity and population differentiation in acoustic signalling. *Functional Ecology*, **28**, 660–668.

Lampe, U., Schmoll, T., Franzke, A. *et al.* (2012) Staying tuned: grasshoppers from noisy roadside habitats produce courtship signals with elevated frequency components. *Functional Ecology*, **26**, 1348–1354.

Lande, R. (1993) Risks of population extinction from demographic and environmental stochasticity and random catastrophes. *The American Naturalist*, **142**, 911–927.

Lättman, H., Bergman, K., Rapp, M. *et al.* (2014) Decline in lichen biodiversity on oak trunks due to urbanization. *Nordic Journal of Botany*, **32**, 518–528.

Leblanc, S.C.F. and DeSloover, J. (1970) Relation between industrialization and the distribution and growth of epiphytic lichens and mosses in Montreal. *Canadian Journal of Botany*, **48**, 1485–1496.

Lenten, L.J.A. and Moosa, I.A. (2003) An empirical investigation into long-term climate change in Australia. *Environmental Modelling and Software*, **18**, 59–70.

Lesage, V., Barrette, C., Kingsley, M.C.S. *et al.* (1999) The effect of vessel noise on the vocal behaviour of belugas in the St. Lawrence River estuary, Canada. *Marine Mammal Science*, **15**, 65–84.

Levins, R. (1969) Some demographic and genetic consequences of environmental heterogeneity for biological control. *Bulletin of the Ecological Society of America*, **15**, 237–240.

Lewis, J.C., Sallee, K.L., and Golightly, R.T. Jr., (1999) Introduction and range expansion of nonnative red foxes (*Vulpes vulpes*) in California. *American Midland Naturalist*, **142**, 372–381.

Lindenmayer, D.B. and Franklin, J.F. (2002) *Conserving Forest Biodiversity: A Comprehensive Multi-scaled Approach*, Island Press, Washington DC.

Longcore, T. and Rich, C. (2004) Ecological light pollution. *Frontiers in Ecology and the Environment*, **2**, 191–198.

Lowe, S., Browne, M., Boudjelas, S. *et al.* (2000) *100 of the World's Worst Invasive Alien Species – A selection from the Global Invasive Species Database*. Invasive Species Specialist Group, Species Survival Commission, International Union for the Conservation of Nature (IUCN): Gland, Switzerland. http://www.issg.org/database/species/reference_files/100English.pdf (accessed 10/10/2013)

Luniak, M. (2004) Synurbization – adaptation of animal wildlife to urban development, in *Proceedings 4th International Symposium on Urban Wildlife Conservation* (eds W.W. Shaw, L.K. Harris, and L. Vandruff), School of Natural Resources, University of Arizona, Tucson, pp. 50–55.

Ma, Z., Gan, X., Choi, C. *et al.* (2007) Wintering bird communities in newly-formed wetland in the Yangtze River estuary. *Ecological Research*, **22**, 115–124.

Maharjan, M.R. (1998) The flow and distribution of costs and benefits in the Chuliban community forest, Dhankuta district, Nepal. *Rural Development Forestry Network, Network Paper 23e*, Summer 1998. http://www.odi.org/sites/odi.org.uk/files/odi-assets/publications-opinion-files/1186.pdf (accessed 18/07/2015)

Marchand, M.N. and Litvaitis, J.A. (2004) Effects of habitat features and landscape composition on the population structure of a common aquatic turtle in a region undergoing rapid development. *Conservation Biology*, **18**, 758–767.

McCarthy, M.A., Lindenmayer, D.B., and Drechsler, M. (1997) Extinction debts and risks faced by abundant species. *Conservation Biology*, **11**, 221–226.

McCauley, R.D., Fewtrell, J., and Popper, A.N. (2003) High intensity anthropogenic sound damages fish ears. *Journal of the Acoustical Society of America*, **113**, 638–642.

McClenaghan, L.R. and Truesdale, H.D. (2002) Genetic structure of endangered Stephens' kangaroo rat populations in southern California. *Southwestern Naturalist*, **47**, 539–549.

McClure, C.J.W., Ware, H.E., Carlisle, J. *et al.* (2013) An experimental investigation into the effects of traffic noise on distributions of birds: avoiding the phantom road. *Proceedings of the Royal Society B* **280**, 20132290. 10.1098/rspb.2013.2290

McDonald, R.I., Kareiva, P., and Forman, R.T.T. (2008) The implications of current and future urbanization for global protected areas and biodiversity conservation. *Conservation Biology*, **141**, 1695–1703.

McKenna, D., McKenna, K., Malcom, S.B. *et al.* (2001) Mortality of lepidoptera along roadways in Central Illinois. *Journal of the Lepidopterists' Society*, **55**, 63–68.

McKinney, M.L. (2002) Urbanization, biodiversity, and conservation. *BioScience*, **52**, 883–890.

Menzel, A. (2003) Plant phenological anomalies in Germany and their relation to air temperature and NAO. *Climate Change*, **57**, 243–263.

Milner-Gulland, E.J., Bennett, E.L. and the SCB 2002 Annual Meeting Wild Meat Group (2003) Wild meat: the bigger picture. *Trends in Ecology and Evolution* **18**, 351-357.

Mitchell, J.C., Brown, R.E.J., and Bartholomew, B. (2008) Urban Herpetology, in *Herpetological Conservation, Number 3*, Society for the Study of Amphibians and Reptiles, Salt Lake City.

Montgomery, J. (1998) Making a city: urbanity, vitality, and urban design. *Journal of Urban Design*, **3**, 93–116.

Morgan, J.W. (1998) Patterns of invasion of an urban remnant of a species-rich grassland in south-eastern Australia by non-native plant species. *Journal of Vegetation Science*, **9**, 181–190.

Morrissey, C.A., Stanton, D.W.G., Tyler, C.R. *et al.* (2014) Developmental impairment in Eurasian dipper nestlings exposed to urban stream pollutants. *Environmental Toxicology and Chemistry*, **33**, 1315–1323.

Muerk, C.D., Zvyagna, N., Gardner, R.O. *et al.* (2009) Environmental, social and spatial determinants of urban arboreal character in Auckland, New Zealand, in *Ecology of Cities and Towns: a Comparative Approach* (eds M.J. McDonnell, A.K. Hahs, and J.H. Breuste), Cambridge University Press, Cambridge, pp. 287–307.

Muñoz, P.T., Torres, F.P., and Megías, A.G. (2015) Effects of roads on insects: a review. *Biodiversity and Conservation*, **24**, 659–682.

Murison, G., Bullock, J.M., Underhill-Day, J. *et al.* (2007) Habitat type determines the effects of disturbance on the breeding productivity of the Dartford Warbler *Sylvia undata*. *Ibis*, **149**, 16–26.

New, T.R. (2007) Politicians, poison, and moths: ambiguity over the icon status of the Bogong moth (*Agrotis infusa*) (Noctuidae) in Australia. *Journal of Insect Conservation*, **11**, 219–220.

Newbound, M., McCarthy, M.A., and Lebel, T. (2010) Fungi and the urban environment: A review. *Landscape and Urban Planning*, **96**, 138–145.

Newbound, M.G. 2008. Fungal diversity in remnant vegetation patches along an urban to rural gradient. PhD Thesis, University of Melbourne, Melbourne.

Noël, S., Ouellet, M., Galois, P. *et al.* (2007) Impact of urban fragmentation on the genetic structure of the eastern red-backed salamander. *Conservation Genetics*, **8**, 599–606.

Nyambod, E.M. (2010) Environmental consequences of rapid urbanisation: Bamenda City, Cameroon. *Journal of Environmental Protection*, **1**, 15–23.

Parks, S.E., Clark, C.W., and Tyack, P.L. (2007) Short and long-term changes in right whale calling behavior: The potential effects of noise on acoustic communication. *Journal of the Acoustical Society of America*, **122**, 3725–3731.

Parris, K.M. (2015) Ecological impacts of road noise and options for mitigation, in *Ecology of Roads: A Practitioners' Guide to Impacts and Mitigation* (eds R. van der Ree, C. Grilo, and D. Smith), Wiley-Blackwell, New York, pp. 151–158.

Parris, K.M. and Hazell, D.L. (2005) Biotic effects of climate change in urban environments: the case of the grey-headed flying-fox (*Pteropus poliocephalus*) in Melbourne, Australia. *Biological Conservation*, **124**, 267–276.

Parris, K.M. and Schneider, A. (2009) Impacts of traffic noise and traffic volume on birds of roadside habitats. *Ecology and Society* **14**, 29. http://www.ecologyandsociety.org/vol14/iss1/art29/

Parris, K.M., M. Velik-Lord, and J.M.A. North. 2009. Frogs call at a higher pitch in traffic noise. *Ecology and Society* **14**, 25. http://www.ecologyandsociety.org/vol14/iss1/art25/

Partecke, J., Van't Hof, T.J., and Gwinner, E. (2005) Underlying physiological control of reproduction in urban and forest-dwelling European blackbirds *Turdus merula*. *Journal of Avian Biology*, **36**, 295–305.

Peacock, D.S., Van Rensburg, B.J., and Robertson, M.P. (2007) The distribution and spread of the invasive alien common myna, *Acridotheres tristis* L. (Aves: Sturnidae), in southern Africa. *South African Journal of Science*, **103**, 465–473.

Pell, A.S. and Tidemann, C.R. (1997a) The ecology of the common Myna in urban nature reserves in the Australian Capital Territory. *EMU*, **97**, 141–149.

Pell, A.S. and Tidemann, C.R. (1997b) The impact of two exotic hollow-nesting birds on two native parrots in savannah and woodland in eastern Australia. *Biological Conservation*, **79**, 145–153.

Peres, C.A. and Palacios, E. (2007) Basin-wide effects of game harvest on vertebrate population densities in Amazonian forests: implications for animal-mediated seed dispersal. *Biotropica*, **39**, 304–315.

Poot, H., Ens, B.J., de Vries, H. *et al.* (2008) Green light for nocturnally migrating birds. *Ecology and Society* **13**, 47. http://www.ecologyandsociety.org/vol13/iss2/art47/

Popper, A.N. and Hastings, M.C. (2009) The effects of anthropogenic sources of sound on fishes. *Journal of Fish Biology*, **75**, 455–489.

Porter, E.E., Forschner, B.R., and Blair, R.B. (2001) Woody vegetation and canopy fragmentation along a forest-to-urban gradient. *Urban Ecosystems*, **5**, 131–151.

Potvin, D.A., Parris, K.M., and Mulder, R.A. (2011) Geographically pervasive effects of urban noise on frequency and syllable rate of songs and calls in silvereyes (*Zosterops lateralis*). *Proceedings of the Royal Society B*, **278**, 2464–2469.

Potvin, D.A. and Mulder, R.A. (2013) Immediate, independent adjustment of call pitch and amplitude in response to varying background noise by silvereyes (*Zosterops lateralis*). *Behavioral Ecology*, **24**, 1363–1368.

Powell, R. and R.W. Henderson. 2008. Urban herpetology in the West Indies. in J.C. Mitchell, R.E. Jung Brown and B. Bartholomew (eds), *Urban Herpetology. Herpetological Conservation vol. 3*, Society for the Study of Amphibians and Reptiles, Salt Lake City, pp. 389–404.

Puckett, K.J., Nieboer, E., Flora, W.P. *et al.* (1973) Sulphur dioxide: its effect on photo-synthetic ^{14}C fixation in lichens and suggested mechanisms of phytotoxicity. *New Phytologist*, **72**, 141–154.

Pulliam, H.R. (1988) Sources, sinks, and population regulation. *American Naturalist*, **132**, 652–661.

Pyke, G.H. (2008) Plague minnow or mosquito fish? A review of the biology and impacts of introduced *Gambusia* species. *Annual Review of Ecology, Evolution, and Systematics*, **39**, 171–191.

Pyle, R.M. (1995) A history of Lepidoptera conservation, with special reference to its Remingtonian debt. *Journal of the Lepidopterists' Society*, **49**, 397–411.

Ranta, P. (2001) Changes in urban lichen diversity after a fall in sulphur dioxide levels in the city of Tampere, SW Finland. *Annales Botanici Fennici*, **38**, 295–304.

Rao, R.S.P. and Girish, M.K.S. (2007) Road kills: assessing insect casualties using flagship taxon. *Current Science*, **92**, 830–837.

Rebele, F. (1994) Ecology and special features of urban ecosystems. *Global Ecology and Biogeography Letters*, **4**, 173–187.

Reed, D.H. and Frankham, R. (2003) Correlation between fitness and genetic diversity. *Conservation Biology*, **17**, 230–237.

Reijnen, R. and Foppen, R. (1994) The effects of car traffic on breeding bird populations in woodland. I. Evidence of reduced habitat quality for Willow Warblers (*Phylloscopus trochilus*) breeding close to a highway. *Journal of Applied Ecology*, **31**, 85–94.

Reijnen, R., Foppen, R., and Meeuwsen, H. (1996) The effects of traffic on the density of breeding birds in Dutch agricultural grasslands. *Biological Conservation*, **75**, 255–260.

Reijnen, R., Foppen, R., ter Braak, C. *et al.* (1995) The effects of car traffic on breeding bird populations in woodland. *III. Reduction of density in relation to the proximity of main roads. Journal of Applied Ecology*, **32**, 187–202.

Resh, V.H. and Jackson, J.K. (1993) Rapid assessment approaches to biomonitoring using benthic macroinvertebrates, in *Freshwater Biomonitoring and Benthic Macroinvertebrates* (eds D.M. Rosenberg and V.H. Resh), Chapman and Hall, New York, pp. 195–233.

Richardson, D.M. and Pyšek, P. (2012) Naturalization of introduced plants: ecological drivers of biogeographical patterns. *New Phytologist*, **196**, 383–396.

Riley, S.J. and Banks, R.G. (1996) The role of phosphorus and heavy metals in the spread of weeds in urban bushland: an example from the Lane Cove Valley, NSW, Australia. *The Science of the Total Environment*, **182**, 39–52.

Riley, S.P.D., Pollinger, J.P., Sauvajot, R.M. *et al.* (2006) A southern California freeway is a physical and social barrier to gene flow in carnivores. *Molecular Ecology*, **15**, 1733–1742.

Riley, S.P.D., Sauvajot, R.M., Fuller, T.K. *et al.* (2003) Effects of urbanization and habitat fragmentation on bobcats and coyotes in southern California. *Conservation Biology*, **17**, 566–576.

Robbins, C.T., Schwartz, C.C., and Felicetti, L.A. (2004) Nutritional ecology of ursids: a review of newer methods and management implications. *Ursus*, **15**, 161–171.

Robinson, N.A. and Marks, C.A. (2001) Genetic structure and dispersal of red foxes (*Vulpes vulpes*) in urban Melbourne. *Australian Journal of Zoology*, **49**, 589–601.

Roetzer, T., Wittenzeller, M., Haeckel, H. *et al.* (2000) Phenology in central Europe – differences and trends of spring phenophases in urban and rural areas. *International Journal of Biometeorology*, **44**, 60–66.

Ross, D.J., Johnson, C.R., and Hewitt, C.L. (2002) Impact of introducing seastars *Asterias amurensis* on survivorship of juvenile commercial bivalves *Fulvia tenuicostata*. *Marine Ecology Progress Series*, **241**, 99–112.

Rubbo, M.J. and Kiesecker, J.M. (2005) Amphibian breeding distribution in an urbanized landscape. *Conservation Biology*, **19**, 504–511.

Ruiz-Avila, R.J. and Klemm, V.V. (1996) Management of *Hydrocotyle ranunculoides* L.f., an aquatic invasive weed of urban waterways in Western Australia. *Hydrobiologia*, **340**, 187–190.

Rydell, J. and Racey, P.A. (1995) Street lamps and the feeding ecology of insectivorous bats. *Symposium of the Zoological Society of London*, **67**, 291–307.

Samuel, Y., Morreale, S.J., Clark, C.W. *et al.* (2005) Underwater, low-frequency noise in a coastal sea turtle habitat. *The Journal of the Acoustical Society of America*, **117**, 1465–1472.

Sattler, T., Borcard, D., Alettaz, R. *et al.* (2010) Spider, bee, and bird communities in cities are shaped by environmental control and high stochasticity. *Ecology*, **91**, 3343–3353.

Sauvajot, R.M., Buechner, M., Kamradt, D.A. *et al.* (1998) Patterns of human disturbance and response by small mammals and birds in chaparral near urban development. *Urban Ecosystems*, **2**, 279–297.

Scheifele, P.M., Andrew, S., Cooper, R.A. *et al.* (2005) Indication of a Lombard vocal response in the St. Lawrence River beluga. *The Journal of Acoustic Society of America*, **117**, 1486–1492.

Schlaepfer, M.A., Runge, M.C., and Sherman, P.W. (2002) Ecological and evolutionary traps. *Trends in Ecology and Evolution*, **17**, 474–480.

Seiler, A., Helldin, J., and Seiler, C. (2004) Road mortality in Swedish mammals: results of a drivers' questionnaire. *Wildlife Biology*, **10**, 225–233.

Sharpe, D.M., Stearns, F., Leitner, L.A. *et al.* (1986) Fate of natural vegetation during urban development of rural landscapes in Southeastern Wisconsin. *Urban Ecology*, **9**, 267–287.

Shannon, G., Angeloni, L.M., Wittemyer, G. *et al.* (2014) Road traffic noise modifies behaviour of a keystone species. *Animal Behaviour*, **94**, 135–141.

Shochat, E., Warren, P.S., and Faeth, S.H. (2006) Future directions in urban ecology. *Trends in Ecology and Evolution*, **21**, 661–662.

Singer, F.J. (1978) Behavior of mountain goats in relation to U.S. Highway 2, Glacier National Park, Montana. *Journal of Wildlife Management*, **42**, 591–597.

Skórka, P., Lenda, M., Moron, D. *et al.* (2013) Factors affecting road mortality and the suitability of road verges for butterflies. *Biological Conservation*, **159**, 148–157.

Slabbekoorn, H. and Peet, M. (2003) Birds sing at a higher pitch in urban noise. *Nature*, **424**, 267.

Slabbekoorn, H., Bouton, N., van Opzeeland, I. *et al.* (2010) A noisy spring: the impact of globally rising underwater sound levels on fish. *Trends in Ecology and Evolution*, **25**, 419–427.

Snodgrass, J.W., Casey, R.E., Joseph, D. *et al.* (2008) Microcosm investigations of stormwater pond sediment toxicity to embryonic and larval amphibians: Variation in sensitivity among species. *Environmental Pollution*, **154**, 291–297.

Soulé, M.E., Alberts, A.C., and Bolger, D.T. (1992) The effects of habitat fragmentation on chaparral plants and vertebrates. *Oikos*, **63**, 39–47.

Soulé, M.E., Bolger, D.T., Alberts, A.C. *et al.* (1988) Reconstructing dynamics of rapid extinction of chaparral-requiring birds in urban habitat island. *Conservation Biology*, **2**, 75–92.

Spinks, P.Q., Pauly, G.B., Crayon, J.J. *et al.* (2003) Survival of the western pond turtle (*Emys marmorata*) in an urban California environment. *Biological Conservation*, **113**, 257–267.

Steinberg, D.A., Pouyat, R.V., Parmelee, R.W. *et al.* (1997) Earthworm abundance and nitrogen mineralization rates along an urban-rural land use gradient. *Soil Biology and Biochemistry*, **29**, 427–430.

Stockin, K.A., Lusseau, D., Binedell, V. *et al.* (2008) Tourism affects the behavioural budget of the common dolphin *Delphinus* sp. in the Hauraki Gulf, New Zealand. *Marine Ecology Progress Series*, **355**, 287–295.

Stoian, D. (2005) Making the best of two worlds: rural and peri-urban livelihood options sustained by non-timber forest products from the Bolivian Amazon. *World Development*, **33**, 1473–1490.

Stone, E.L., Jones, G., and Harris, S. (2009) Street lighting disturbs commuting bats. *Current Biology*, **19**, 1123–1127.

Storper, M. and Manville, M. (2006) Behaviour, preferences, and cities: urban theory and urban resurgence. *Urban Studies*, **43**, 1247–1274.

Sun, J.W.C. and Narins, P.M. (2005) Anthropogenic sounds differentially affect amphibian call rate. *Biological Conservation*, **121**, 419–427.

Temple, S.A. (1987) Predation on turtle nests increases near ecological edges. *Copeia*, **1987**, 250–252.

Thompson, B. (2015) Recreational trails reduce the density of ground-dwelling birds in protected areas. *Environmental Management*, **55**, 1181–1190.

Thompson, K., Austin, K.C., Smith, R.M. *et al.* (2003) Urban domestic gardens (I): putting small-scale plant diversity in context. *Journal of Vegetation Science*, **14**, 71–78.

Thompson, K. and Jones, A. (1999) Human population density and prediction on local plant extinction in Britain. *Conservation Biology*, **13**, 185–189.

Tilman, D., May, R.M., Lehman, C.L. *et al.* (1994) Habitat destruction and the extinction debt. *Nature*, **371**, 65–66.

Torok, S.J., Morris, C.J.G., Skinner, C. *et al.* (2001) Urban heat island features of southeast Australian towns. *Australian Meteorological Magazine*, **50**, 1–13.

Torok, S.J. and Nicholls, N. (1996) A historical annual temperature dataset for Australia. *Australian Meteorological Magazine*, **45**, 251–260.

Trombulak, S.C. and Frissell, C.A. (2000) Review of ecological effects of roads on terrestrial and aquatic communities. *Conservation Biology*, **14**, 18–30.

Underhill-Day, J.C. and Liley, D. (2007) Visitor patterns on southern heaths: a review of visitor access patterns to heathlands in the UK and the relevance to Annex I bird species. *Ibis*, **149**, 112–119.

van der Ree, R., McDonnell, M.J., Temby, I.D. *et al.* (2006) The establishment and dynamics of a recently established urban camp of Pteropus poliocephalus outside their geographic range. *Journal of Zoology*, **268**, 177–185.

van Heezik, Y., Smyth, A., Adams, A. *et al.* (2010) Do domestic cats impose an unsustainable harvest on urban bird populations? *Biological Conservation*, **143**, 121–130.

van Langevelde, F., Ettema, J.A., Donners, M. *et al.* (2011) Effects of spectral composition of artificial light on the attraction of moths. *Biological Conservation*, **144**, 2274–2281.

Vandergast, A.G., Bohonak, A.J., Weissman, D.B. *et al.* (2007) Understanding the genetic effects of recent habitat fragmentation in the context of evolutionary history: phylogeography and landscape genetics of a southern California endemic Jerusalem cricket (Orthoptera: Stenopelmatidae: *Stenopelmatus*). *Molecular Ecology*, **16**, 977–992.

Vasconcelos, R.O., Amorim, M.C.P., and Ladich, F. (2007) Effects of ship noise on the detectability of communication signals in the Lusitanian toadfish. *Journal of Experimental Biology*, **210**, 2104–2112.

Vasconcelos, R.O., Carriço, R., Ramoa, A. *et al.* (2012) Vocal behavior predicts reproductive success in a teleost fish. *Behavioral Ecology*, **23**, 375–383.

Vasconcelos, R.O., Simões, J.M., Almada, V.C. *et al.* (2010) Vocal behaviour during territorial intrusions in the Lusitanian toadfish: boatwhistles also function as territorial 'keep-out' signals. *Ethology*, **116**, 155–165.

Walsh, C.J., Roy, A.H., Feminella, J.W. *et al.* (2005) The urban stream syndrome: current knowledge and the search for a cure. *Journal of North American Benthological Society*, **24**, 706–723.

Walsh, C.J., Waller, K.A., Gehling, J. *et al.* (2007) Riverine invertebrate assemblages are degraded more by catchment urbanisation than by riparian deforestation. *Freshwater Biology*, **52**, 574–587.

Warren, P.S., Katti, M., Ermann, M. *et al.* (2006) Urban bioacoustics: it's not just noise. *Animal Behaviour*, **71**, 491–502.

Webb, J.K. and Shine, R. (2000) Paving the way for habitat restoration: can artificial rocks restore degraded habitats of endangered reptiles? *Biological Conservation*, **92**, 93–99.

Wellborn, G.A., Skelly, D.K., and Werner, E.E. (1996) Mechanisms creating community structure across a freshwater habitat gradient. *Annual Review of Ecology and Systematics*, **27**, 337–363.

White, M.A., Nemani, R.R., Thornton, P.E. *et al.* (2002) Satellite evidence of phenological differences between urbanized and rural areas of the Eastern United States broadleaf forest. *Ecosystems*, **5**, 260–277.

Williams, N.S.G., Schwartz, M.W., Vesk, P.A. *et al.* (2009) A conceptual framework for predicting the effects of urban environments on floras. *Journal of Ecology*, **97**, 4–9.

Williams, N.S.G., Morgan, J.W., McDonnell, M.J. *et al.* (2005) Plant traits and local extinctions in natural grasslands along an urban-rural gradient. *Journal of Ecology*, **93**, 1203–1213.

Williams, R., Lusseau, D., and Hammond, P.S. (2006) Estimating relative energetic costs of human disturbance to killer whales (*Orcinus orca*). *Biological Conservation*, **133**, 301–311.

Witherington, B.E. and Bjorndal, K.A. (1991) Influences of artificial lighting on the seaward orientation of hatchling Loggerhead Turtles *Caretta caretta*. *Biological Conservation*, **55**, 139–149.

Wood, W.E. and Yezerinac, S.M. (2006) Song Sparrow (*Melospiza melodia*) song varies with urban noise. *The Auk*, **123**, 650–659.

Woods, M., McDonald, R.A., and Harris, S. (2003) Predation of wildlife by domestic cats *Felis catus* in Great Britain. *Mammal Review*, **33**, 174–188.

Yule, C.M., Gan, J.Y., Jinggut, T. *et al.* (2015) Urbanization affects food webs and leaf-litter decomposition in a tropical stream in Malaysia. *Freshwater Science*, **34**, 702–715.

Zedler, J.B. and Kercher, S. (2004) Causes and consequences of invasive plants in wetlands: opportunities, opportunists, and outcomes. *Critical Reviews in Plant Sciences*, **23**, 431–452.

Zhou, L., Dickinson, R.E., Tian, Y. *et al.* (2004) Evidence for a significant urbanization effect on climate in China. *Proceedings of the National Academy of Sciences of the United States of America*, **101**, 9640–9544.

Zipperer, W.C. and Guntenspergen, G.R. (2009) Vegetation composition and structure of forest patches along urban-rural gradients, in *Ecology of Cities and Towns: A Comparative Approach* (eds M.J. McDonnell, A.K. Hahs, and J.H. Breuste), Cambridge University Press, Cambridge, pp. 274–286.

CHAPTER 4

Community-level responses to urbanization

4.1 Introduction

Let's now turn our attention to the question of how ecological communities respond to urbanization. When a town or city is first built, the resident ecological communities can sometimes be destroyed in their entirety. The greater the speed, intensity and/or scale of urban development, the more complete this destruction is likely to be. However, populations of some species will survive the initial phase of urbanization, to be joined over time by populations of additional species (both native and introduced) via dispersal and speciation. Therefore, following the immediate, community-level impacts of habitat loss, effects of urbanization can generally be observed at the level of populations or species. Yet the diverse responses of populations and species to the processes of urbanization can lead to important changes and emerging patterns at the level of the ecological community. The particular characteristics of urban habitats can result in the formation of novel ecological communities, some of which have no obvious analogues in natural environments. Urbanization also has significant effects on terrestrial, freshwater and marine communities beyond the geographical boundaries of a town or city.

Here, I define an ecological community as a group of potentially interacting species that occur together in space and time. This is known as the individualistic community concept (Gleason 1917, 1926). It can be contrasted with the organismal community concept (Clements 1916, 1936), which emphasises interactions between species that have coexisted as a group for an extended period of time, such that the community they form is regarded as a super "organism". The individualistic concept of the community allows us to consider ecological communities across any spatial or temporal scale (Vellend 2010); for example, from the community of invertebrates living in a single tree in a city park or the community of plants living in the cracks of a brick wall, to the community of invertebrates or plants living in an entire city or in every city in the world. Each community can be studied over time scales ranging from a single day to hundreds or thousands of years.

Community ecologists are interested in the diversity, abundance and composition of species within ecological communities, and the underlying processes and

Ecology of Urban Environments, First Edition. Kirsten M. Parris.

interactions that influence community structure (see Box 4.1). The theoretical basis for the coexistence of species and the patterns of species richness and abundance observed in ecological communities has received much attention over the last 50 years, leading to a vast – and often confusing – array of concepts, hypotheses and theories (Palmer 1994; Vellend 2010). In reaction to this proliferation, a number of ecologists have argued that community ecology lacks a sufficiently general theory that applies in all (or at least most) ecological systems and circumstances (e.g., Palmer 1994; Lawton 1999; Simberloff 2004). Vellend (2010) provided a framework for classifying existing community theories with respect to their emphasis on four community-level processes – selection, ecological drift, dispersal and speciation – analogous to the four processes of selection, genetic drift, gene flow and mutation within population genetics.

Box 4.1 Measuring Ecological Communities

Ecologists describe communities in a range of ways, such as by the number of species they contain (species richness), the identity of those species (composition) and how these change over space and time (e.g., beta diversity, phylogenetic diversity, stability, turnover, succession and resilience). The relative abundance of the different species within a community – that is, the proportion of individuals in the community that belongs to each species – can provide additional information about community composition which can summarized using metrics of diversity or evenness, such as the Shannon–Wiener diversity index (Begossi 1996; Spellerberg and Fedor 2003). A range of theories and models has been used to predict the relative abundance of species in ecological communities in non-urban environments, including the lognormal distribution (Preston 1948), the broken-stick model (MacArthur 1960), and the zero-sum multinomial distribution (Hubbell 2001). It is interesting to consider whether the types of community patterns observed in urban areas are similar to those seen in more natural areas. Do the various models of the relative abundance of species in ecological communities make robust predictions in all environments?

A focus on the species richness of ecological communities in urban environments can exclude important information about the particular species that occur there. For example, a suburban garden may support more plant species than a comparable area of native grassland, woodland or desert on the city's fringe (see Section 3.3.1). However, if the species in the garden include many common, exotic, horticultural species that were planted by the owners of the garden, while the native vegetation community includes rare, endemic species, then the ecological and conservation value of the former could be considered as less than those of the latter. On a larger spatial scale, vascular plant communities in urban areas across Germany have higher species richness than those in agricultural or semi-natural areas, but this is not matched by a higher phylogenetic diversity (Knapp et al. 2008). Much of the species richness of urban plant communities is due to a large number of closely related, functionally similar species that thrive in urban areas.

The composition of ecological communities is often influenced by interactions between species. These include competition for resources, such as space, nutrients, water and light (in the case of fungi and plants) or food, water, shelter and breeding sites (in the case of animals). Other inter-specific interactions that can shape the composition of communities include predation, mutualism, parasitism and predator-mediated competition (Polis et al. 1989; Hatcher et al. 2006). As discussed in Chapter 3, urbanization can alter the availability of resources for organisms over space and time in quite dramatic ways. These alterations subsequently affect the interactions between species, leading to changes in the diversity and composition of ecological communities. The uniqueness or generality of the community-level changes observed in different cities around the world is a key question in urban ecology.

In community ecology, selection arises from inherent differences between species (species-level traits) and between individuals of the same species (intraspecific variation) that influence their fitness in a given environment (Vellend 2010; Bolnick et al. 2011; Violle et al. 2012). Ecological drift results from stochasticity or randomness within communities; dispersal is the movement of individuals through geographic space; and speciation is the evolution of new species over time (Vellend 2010). For example, niche theories emphasize selection, simple neutral theory emphasizes ecological drift, and metacommunity theory considers selection, drift and dispersal. Vellend's (2010) general theory of ecological communities considers all four processes; species are added to communities via speciation and dispersal, and their relative abundances are then shaped by ecological drift, selection and further dispersal. Nemergut et al. (2013) proposed a slight modification to Vellend's scheme, which uses evolutionary diversification rather than speciation as the fourth community-level process. This broadens the consideration of evolutionary change and its potential impacts on community dynamics beyond the particular situation in which new species are created. It also includes important evolutionary changes in organisms for which the traditional species concept may be inappropriate, such as many microbes (Konstantinidis et al. 2006; Burke et al. 2011; Nemergut et al. 2013).

Whether general or specific, few theories in community ecology have been developed with urban environments in mind. So, to what extent do these theories apply in urban areas? Can they be used to help understand the impacts of urbanization at the level of the ecological community? Can they predict characteristics of urban ecological communities such as species richness, composition, relative abundance, or spatial and temporal dynamics? Or do we need a new ecological theory for urban communities? In this chapter, I address these questions while examining some of the key community-level effects of urbanization. One way to think about the impacts of urbanization on ecological communities is to consider how urbanization alters the four community-level processes of selection, ecological drift, dispersal and evolutionary diversification; I take this approach here.

4.2 Selection: niche theories in urban ecology

4.2.1 The ecological niche and the environmental gradient

At the level of the ecological community, the process of selection depends on functional differences between species and between individuals of the same species. We can apply the concept of selection to the species in a community rather than the alleles in a population by considering the "population" to be all the individuals of all the species within a community, and (in the simple case) the species to which an individual belongs as the characteristic of primary interest. Just as selection may favour one allele over another in a population, selection may favour individuals of one species over individuals of another in a community (Vellend 2010). These differences influence the way in which each species interacts with its environment and with the other species that it encounters. Selection can be constant (when the fitness of individuals within a species does not change, and fitness differs between species in a consistent way) or density dependent (when the fitness of individuals within a species depends on the density of individuals of that species and/or other species in the community). Whether constant or density-dependent, selection can also vary through space and time (Vellend 2010).

For example, under particular environmental conditions, individuals of one species in the available pool may have a fitness advantage over individuals of others – that is, they are more likely to survive and reproduce. When environmental conditions vary across geographic space, different species are favoured in different locations. Observations of these kinds of patterns in nature led to the concept of the ecological niche. The term niche was first used in ecology by Grinnell (1917), but was later defined by Hutchinson (1957) as the multi-dimensional environmental space (or hypervolume) in which a species could exist. Each dimension of this space represents an environmental variable or resource that is relevant to the species in question, considered as a continuum or gradient. For example, let's consider two environmental variables – annual mean air temperature and annual mean rainfall. One species of tree might thrive in cool, moist conditions, while another thrives in hotter, drier conditions. The two tree species occupy a different part of the environmental space defined by the two climatic variables, and the abundance of each species is expected to vary along each climatic gradient (Figure 4.1). The environmental space that delimits the niche of a species gains an additional dimension for each additional environmental variable that is considered (e.g., soil depth, aspect, solar radiation).

The entire environmental space that a species could occupy in the absence of other species is known as its fundamental niche. However, the actual environmental space that a species occupies in nature – its realized niche – is often much smaller than this, because individuals of multiple species may be seeking to use the same resources at a given time and location. As the overlap between the fundamental niches of sympatric species increases, we would expect the interaction between them also to increase. The competitive exclusion principle (also known as Gause's principle; Gause 1934; Hardin 1960) states that two species with identical

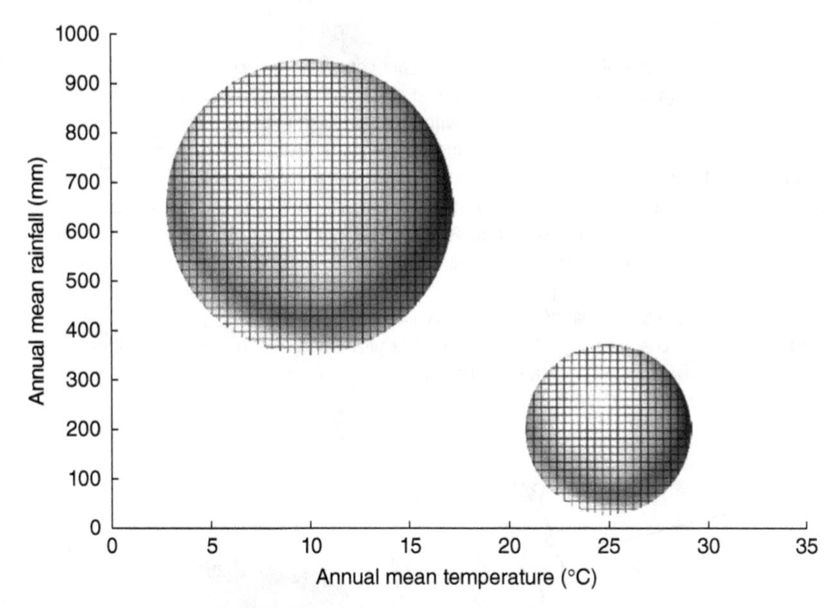

Figure 4.1 Schematic diagram of the ecological niche of two tree species defined by two environmental variables, annual mean air temperature and annual mean rainfall; the abundance of each species (demonstrated by a 3D dome) increases from the edges to the centre of its niche. The species on the left is adapted to cool, moist conditions and has a larger ecological niche than the species on the right, which is adapted to hotter, drier conditions.

fundamental niches cannot coexist indefinitely, as one species must eventually exclude the other via displacement or by causing the local extinction of the other. While the principle has rarely been observed to hold in natural, stochastic environments, it provides an important theoretical building block for both descriptive and quantitative community ecology. The niche concept is central to a range of other concepts, theories and models in community ecology, such as the gradient paradigm, all types of habitat models (including those incorporating functional traits), the ecological-guild concept, resource-competition models, predator-prey models, food webs and trophic cascades (Box 4.2).

Box 4.2 Food Webs and Trophic Interactions in Urban Habitats

Food webs map the trophic interactions between species in an ecological community. The most commonly considered type of trophic interaction is that of who eats whom in the community (hence the term food web), but trophic interactions also include who is a parasite or parasitoid of whom, who competes with whom, and who is in a mutualistic relationship with whom. Individual species, or groups of species that have the same predators and prey (known as functional groups), form nodes within a food web; nodes are connected by trophic links that represent the flow of energy between them (Thompson et al. 2012; Figure 4.2). Food webs have a number of trophic levels or positions, starting with saprotrophs (microorganisms that break down organic material) and autotrophs (organisms that generate their own food, such as plants) at the base, followed by herbivores (animals that eat plants), primary predators (animals that eat herbivores), secondary predators (animals that eat primary predators) and omnivores (animals that eat both plants and animals). Organisms in these last four groups are all heterotrophs, so-called because they consume other organisms for food. The highest trophic level of a food web contains the top or apex predator(s), which have no predators of their own (Figure 4.2), although they often have various parasites. While food webs can be very complex, they tend to share certain characteristics. These include the proportion of top predators, intermediate species and basal species, and the number of trophic levels in the web (typically 3–5; Pimm et al. 1991).

Figure 4.2 A simplified urban food web showing (1) saprotrophs (fungi), (2) autotrophs (plants), (3) herbivores or primary consumers (ant, dove, moth), (4) primary predators (blackbird, insectivorous bat) and (5) secondary predators (cat, fox).

The abundance and activity of top predators appear to strongly influence the structure and function of some food webs (top-down control). In contrast, the abundance of the basal species or the input of nutrients into the system appears to regulate other food webs (bottom-up control). In practice, most food webs are likely to contain an element of top-down and bottom-up control (Pace et al. 1999). Examples of top-down control can be found in both aquatic and terrestrial food webs, where top predators control the abundance of their prey species (primary predators and/or herbivores), which in turn controls levels of herbivory and the abundance and diversity of basal species (Chase et al. 2000). Frequency-dependent predation, in which individuals of the most abundant species are more likely to be eaten than those belonging to less abundant species, can also act as a stabilizing mechanism within ecological communities, reducing fitness differences between prey species that may lead to competitive exclusion (Chesson 2000; see Section 4.2.3).

A trophic cascade describes the situation where strong predator–prey effects alter the abundance, biomass or productivity of a species, functional group or trophic level across more than one link in a food web (Pace et al 1999). While trophic cascades often start at the highest trophic level of a food web (i.e., with the top predators), they can also start at lower levels. The processes of urbanization can substantially alter the abundance of species at different trophic levels, and consequently the composition, structure and function of terrestrial and aquatic food webs (reviewed by Faeth et al. 2005). Urbanization can therefore disrupt existing trophic cascades by reducing the abundance of top predators such as large mammalian carnivores, which then leads to an increase in the abundance of secondary, medium-sized predators (also known as mesopredators; Ritchie and Johnson 2009) and a consequent decrease in populations of primary predators and herbivores. At the other end of the food web, changes to water and nutrient inputs in urban habitats can lead to a rapid increase in primary productivity and a possible strengthening of bottom-up effects on herbivore populations, while increased food resources may alter competitive and predatory interactions across a number of trophic levels (Faeth et al 2005; Ritchie and Johnson 2009).

The gradient paradigm (Whittaker 1967) considers environmental (abiotic) variation to be ordered in space, often with a gradual change in a given environmental variable when moving in a particular direction. The resulting spatial patterns of environmental variation then influence the spatial distribution of populations and species, and the composition of communities (e.g., Terborgh 1971; Austin 1987; Buckley and Jetz 2010; Hoverman et al. 2011). Examples of environmental gradients include temperature and rainfall gradients, which in turn can be related to gradients in altitude and latitude (e.g., Bennett et al. 1991; Weaver 2000; Potapova and Charles 2002; Hu et al. 2010). McDonnell and Pickett (1990) first presented the concept of the urban-rural gradient, which considers urban and rural to be at opposite ends of a gradient of human impact upon the earth and its ecosystems. Considering "urban-ness" as a gradient analogous to other environmental gradients rather than as a simple, binary classification (urban or not) represented an important advance in urban ecology. The urban–rural gradient is often a complex one, representing a number of environmental variables that vary across space when moving from urban to rural environments (McDonnell and Pickett 1990; Hahs and McDonnell 2006). Examples include the density and height of built structures, the proportional cover of impermeable surfaces, the concentration of various atmospheric pollutants, and the density of roads. The urban–rural gradient can also be characterized as a temperature gradient, with air temperature decreasing as one moves from the urban to the rural end of the gradient (see Chapter 2; Howard 1833; Bridgman et al. 1995; Torok et al. 2001). In addition, all these environmental variables are highly correlated with the density of the human population in a city or urban region.

Urban–rural gradients have been used as a framework for a wide range of empirical studies of ecological communities in urban habitats. Early conceptualizations of the urban–rural gradient tended to be linear, describing a highly urbanized "core" or central business district surrounded by concentric rings of decreasing urbanization through the inner suburbs, outer suburbs and the urban–rural fringe (e.g., McDonnell et al. 1993; Luck and Wu 2002). As a consequence, some authors have criticized urban–rural gradients as an oversimplification of the spatial complexity of urban areas (e.g., Alberti 2008; Ramalho and Hobbs 2012). However, urban–rural gradients are not necessarily linear. Instead, they can take a number of forms, depending on the intensity and spatial pattern of urbanization throughout a particular city or geographic area. The distribution of populations and species, the processes influencing this distribution, and the resulting spatial variation in the composition of communities can all be studied across urban–rural gradients as they are across other types of environmental gradients (e.g., Blair 1996; Niemelä et al. 2002; Lawson et al. 2008; Stracey and Robinson 2012; Hamer and Parris 2013).

A review by McDonnell and Hahs (2008) found more than 200 studies of various taxonomic groups across urban–rural gradients, including birds, insects, mammals, fish, plants and fungi. As observed along other environmental gradients, different species or guilds within each taxonomic group respond to urban–rural gradients in different ways. For example, a study of bird communities across a gradient of urbanization in Pretoria, South Africa, found a total of 65 native species in semi-natural habitats (at the rural end of the gradient), compared to 45 native species in suburban habitats and 47 in urban habitats (van Rensburg et al. 2009). The species lost with urbanization tended to be grassland specialists, such as the African stonechat *Saxicola torquatus*, the rufous-naped lark *Mirafra africana* and the African quailfinch *Ortygospiza atricollis* (Figure 4.3). However, other native species that were absent from semi-natural habitats occurred in moderate abundance in suburban and urban areas. This group included the Karoo thrush *Turdus smithi*, the bronze mannikin *Spermestes cucullatus* and the white-bellied sunbird *Cinnyris talatala* (Figure 4.3). Non-native species, such as the rock dove *Columba livia* and house sparrow *Passer domesticus*, were largely absent from semi-natural areas but abundant in urban habitats; species such as these are sometimes referred to as synanthropes or urban exploiters (see Box 7.1 for more information). Overall, the total abundance of birds increased when moving from the rural to the urban end of the gradient (van Rensburg et al. 2009).

Figure 4.3 (a) Pretoria, South Africa, (b) the rufous-naped lark *Mirafra africana*, a grassland specialist lost from Pretoria following urbanization, (c) the white-bellied sunbird *Cinnyris talatala*, which is found in suburban and urban areas of the city. Pictures have been cropped and converted to black and white. (a) Photograph by Petrus Potgieter. (bc) Photographs by Derek Keats. Used under CC-BY-2.0 https://creativecommons.org/licenses/by/2.0/.

These findings are similar to those from other cities around the world; bird communities at the urban end of urban–rural gradients tend to have lower species richness but a higher density of individuals, supported by a relatively plentiful and predictable supply of food and water (e.g., Marzluff 2001; Shochat 2004; Shochat et al. 2004, 2010; also see review by Chace and Walsh 2006; Fig. 4.4). Other common, community-level patterns observed across urban–rural gradients (reviewed by McKinney 2008) include a consistent decline in species richness with increasing urbanization (e.g., 11 studies on amphibians in Europe, North America, South America and Australia); a peak in species richness at an intermediate point along the gradient, often where there is an interface between native vegetation and suburban development leading to increased habitat diversity (e.g., six studies on plants in Europe and North America); and an abrupt decline in species richness at a particular threshold of urbanization (e.g., riverine macroinvertebrates in Melbourne, Australia; Blair and Launer 1997; Walsh et al. 2005, 2007; Sadler et al. 2006). It appears that the scale of analysis can strongly influence the observed patterns of species richness along urban–rural gradients, and this should be considered when designing or interpreting a study (McKinney 2008). Furthermore, some studies of ecological communities along urban–rural gradients have suffered from insufficient replication at the level of the question (that is, the position of the study site on the gradient), which may limit their inferential power (e.g., Blair 1996, 1999).

4.2.2 Habitat models

Habitat models are conceptual or statistical models that relate a response variable (such as the probability of occurrence or abundance of a species, or the species richness of an ecological community) to one or more environmental variables (see review by Wintle et al. 2005). Habitat models help ecologists to identify and formalize important niche relationships (Elith and Leathwick 2009). They may also be used to predict the location of suitable habitat for a species or community by estimating the probability that it will occur in different parts of the landscape, an important application in conservation planning (e.g., Ferrier et al. 2002a, b). Even if it were possible to conduct field surveys for every species of interest in every location across a given area, it would be very expensive and time-consuming to do so. Well-constructed, reliable, and defensible habitat models reduce the need for comprehensive field-data sets when planning the location of different land uses, such as conservation reserves, timber harvesting or urban development.

Conceptual habitat models (known as habitat suitability indices) can be constructed with little or no field data, relying instead on expert opinion to identify relationships between the species or community of interest and environmental variables that influence its distribution (Burgman et al. 2001). However, the quality of these models largely depends on the quality of the expert opinion used to construct them, and a lack of independent field data can make them very difficult to evaluate and defend (Wintle et al. 2005). Where sufficient data are available on

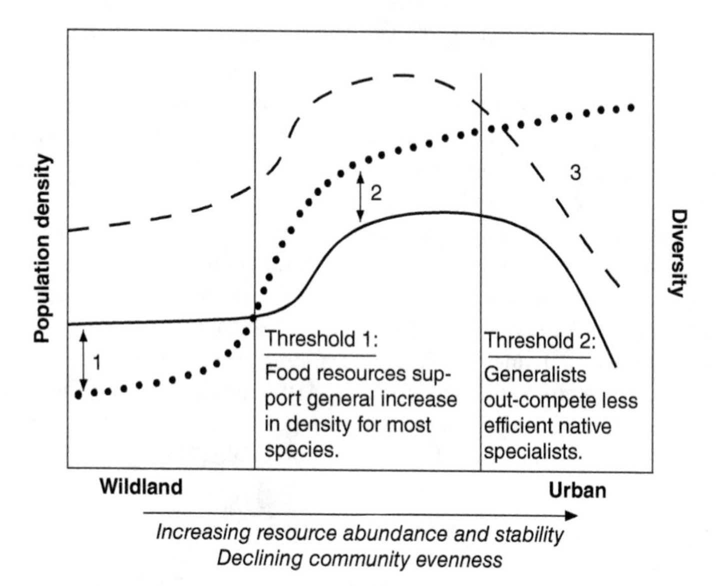

Figure 4.4 Schematic diagram of common patterns observed in urban bird communities; changes in population densities (solid and dotted lines) and species diversity (dashed line) along a wildland-urban gradient. 1 In wildlands, native species (solid line) out-compete invasive, synanthropic species (dotted line). 2 At intermediate levels of urbanization, both groups of species are still present. Invasive species increase in density more rapidly than native species by exploiting novel, stable, and homogenized food resources (e.g., commercial seed mixture). Native species persist through spatial partitioning, exploiting remnant patches that still contain specialized food or other resources, but may decline in the longer term. 3 At high levels of urbanization, coexistence mechanisms collapse, predation pressure relaxes, invasive species become dominant foragers and specialist native species become locally extinct. Shochat et al. 2010, figure 5b. Reproduced with permission of Oxford University Press.

the spatial distribution of a species or community and one or more relevant environmental variables, ecologists can construct statistical habitat models. Examples of these include models that are suitable for unplanned, presence-only data such as BIOCLIM, DOMAIN and MAXENT (Busby 1991; Carpenter et al. 1993; Phillips et al. 2006; also see review by Elith et al. 2006).

Generalized linear models, such as logistic regression and Poisson regression models, generally require information on the presence and absence of each species or community across space – the kind of data collected during systematic surveys (Wintle et al. 2005). Logistic regression models use binary, presence/absence data to estimate the probability of occurrence of a species or community, while Poisson regression models (a special case of the negative binomial) can use count data to estimate the abundance of a species or the species richness of a community as a function of environmental variables (e.g., Parris 2001; Parris 2006; Hamer and Parris 2011). Multivariate association (or ordination) methods, such as principal components analysis and canonical correspondence analysis, relate the presence/absence or abundance of multiple species in a community to environmental variables (Ter Braak 1986, 1987; Ter Braak and Prentice 1988). Partial ordination methods, such as partial redundancy analysis, calculate the amount of variation in the composition of communities that is explained by environmental, spatial, and spatially-structured environmental variables (Borcard et al. 1992; Cottenie 2005; Peres-Neto and Kembel 2015).

Habitat models have been applied to a variety of questions in urban ecology, including the effect of landscape-scale environmental variables on the probability of occurrence of Japanese hares *Lepus brachyurus* along an urban–rural gradient (Saito and Koike 2009), the predicted distribution of the endangered black-headed dwarf chameleon *Bradypodion melanocephalum* in a rapidly urbanizing area of KwaZulu-Natal, South Africa (Armstrong 2009), and the influence of species-level traits on the probability of persistence or local extinction of >8000 species of plants across 11 cities (Duncan et al. 2011; Figure 4.5). Poisson

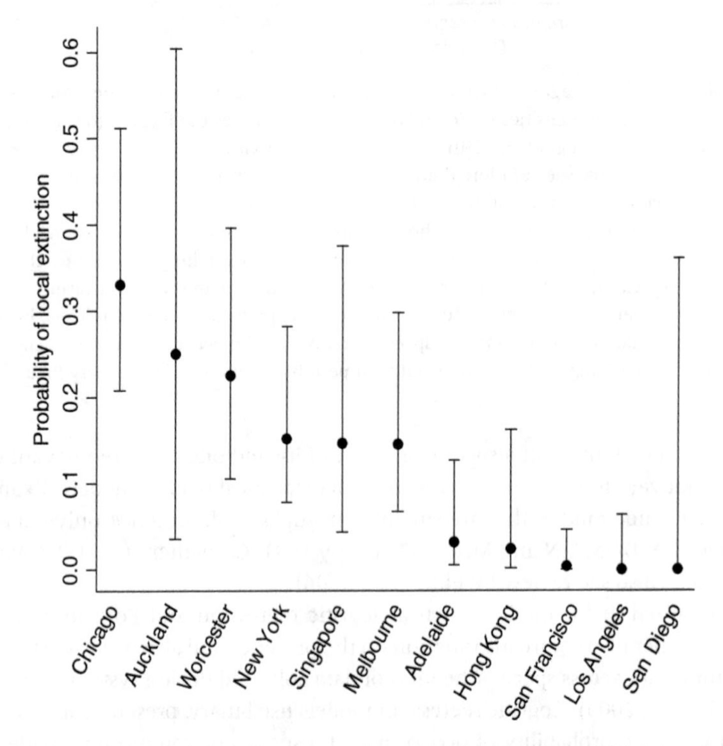

Figure 4.5 Estimated probability of local extinction (circles) with 95% credible intervals for a plant species with traits fixed at the reference class for each categorical trait (herbaceous, annual, non-clonal, abiotic pollination, no specialised dispersal, nitrogen fixing, C3 photosynthetic pathway, no spines), and having mean height and seed mass, for each of 11 cities. The graph shows strong variation in the probability of local extinction between cities. Duncan et al. 2011, Figure 3. Reproduced with permission of John Wiley & Sons.

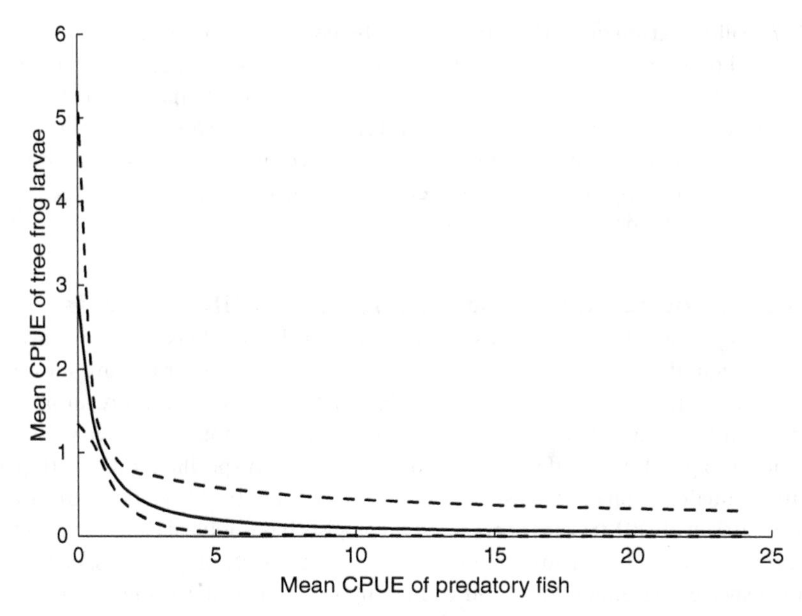

Figure 4.6 Relationship between the abundance of tree frog (*Litoria* spp.) larvae (tadpoles) and the abundance of predatory fish in urban wetlands across Melbourne, Australia, both expressed as catch per unit effort (CPUE). The solid line shows the relationship predicted by a Bayesian, zero-inflated negative binomial regression model (ZINB), and the dotted lines show the 95% credible intervals. Adapted from Hamer & Parris 2013, figure 4a. Reproduced with permission of Springer Science + Business Media.

regression modelling demonstrated a strong negative relationship between the abundance of tree frog (*Litoria* spp.) tadpoles and the abundance of predatory fish in urban wetlands across Melbourne, Australia (Figure 4.6). In an application of habitat modelling to a question of human health in urban environments, both logistic and Poisson regression models were used to identify environmental variables associated with the larval habitat of *Anopheles* mosquitoes in urban Kisumu and Malindi, Kenya (Jacob et al. 2005). Mosquitoes of the genus *Anopheles* are vectors for the malaria parasite, and urban development and expansion is likely to change the nature and distribution of potential habitat for their larvae (Jacob et al. 2005).

Habitat models can also be used to identify environmental variables that influence the persistence of certain habitat types or ecological communities in urbanizing landscapes. Case studies include the probability that farmland on the urban fringe of Beijing, China was converted to residential or industrial land (Liu et al. 2013), and the probability that patches of endangered grassland were lost to urban development or degraded by invasion of non-native grasses on the fringes of Melbourne, Australia (Williams et al. 2005). In the first example, more than 5000 ha of farmland were lost to urban development in the study area between 2004 and

2007. Fallow farmland was more likely to be converted to residential or industrial land than farmland currently in cultivation for crops or vegetables (Liu et al. 2013). In the second example, 1670 ha or 23% of temperate native grasslands in the study area were destroyed by urban development between 1985 and 2000, despite their status as an endangered vegetation community. In both examples, spatial proximity to existing development and/or major roads increased the probability of conversion to urban land uses.

4.2.3 Ecological guilds and resource-competition models

An ecological guild is a group of sympatric species that exploits the same class of environmental resources (e.g., a certain type of food, shelter or nesting site) in a similar way (Root 1967; Simberloff and Dayan 1991). Rather than regarding all species in an ecological community as potential competitors, dividing sympatric species into guilds is an effective way to concentrate on specific groups with particular functional relationships (Simberloff and Dayan 1991; Williams and Hero 1998; Luck et al. 2013; Gates et al. 2015). Ecologists have applied the guild concept to communities in urban environments to address a variety of questions, including which species are exploiting – and competing for – particular resources in cities; which species are most likely to be affected when specific resources are altered or lost through the processes of urbanization; and how the availability of resources is affected when one or more species that previously exploited these resources become locally extinct (e.g., Lim and Sodhi 2004; Pauw and Louw 2012; Hironaka and Koike 2013; Luck et al. 2013; Huijbers et al. 2015).

For example, a study of avian guilds in Singapore found that insectivorous and carnivorous birds (the latter group feeding largely on animals other than insects) were adversely affected by urbanization, while frugivores thrived in areas of low-density housing where there was an abundance of planted fruit trees (Lim and Sodhi 2004). Shrub-nesting birds and those that excavate tree cavities for nesting (such as woodpeckers and kingfishers) also declined with increasing levels of urbanization. Microchiropteran (insectivorous) bats adapted to foraging in open and edge habitats were more abundant across 18 cities and towns in south-eastern Australia than those adapted to foraging in cluttered habitats (Luck et al. 2013; Figure 4.7). Clutter-adapted species mostly forage in areas of dense vegetation; this guild appears to be the most sensitive to urbanization.

When two or more species with overlapping fundamental niches occupy the same geographic area, they will often compete for resources, such as food, water, shelter or breeding sites (mobile animals); germination sites, light, soil moisture or nutrients (plants); settlement sites, light or food (sessile animals with a mobile juvenile stage, such as some marine invertebrates). This competition can have an important influence on the actual distribution – or realized niche – of each species in that geographic area. As outlined in Section 4.2.1, the competitive exclusion

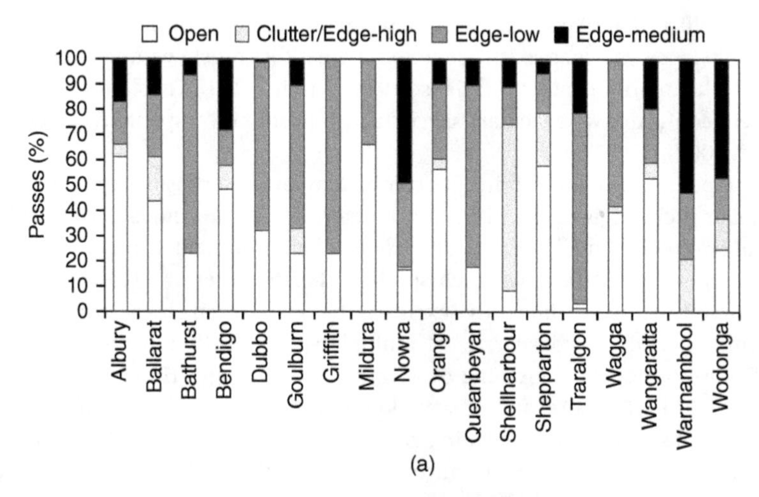

(a)

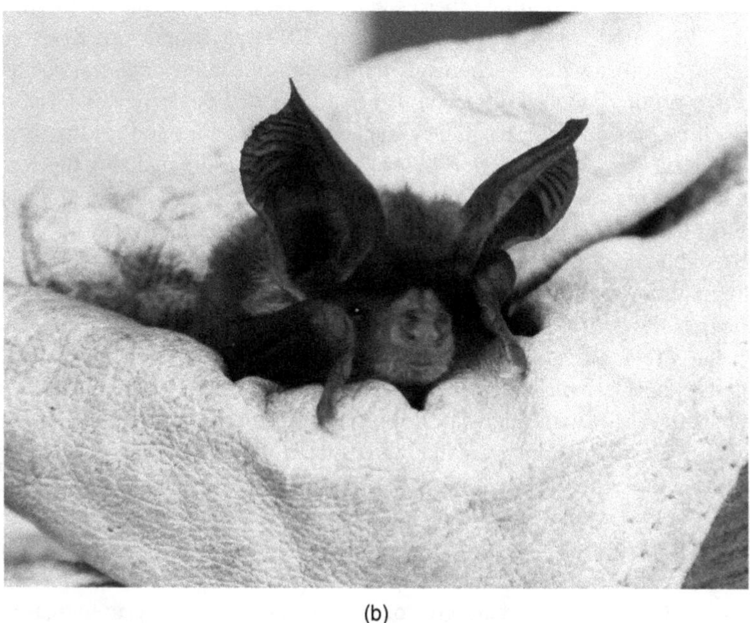

(b)

Figure 4.7 (a) Percentage of passes in each of 18 towns and cities in south-eastern Australia by bats belonging to five functional guilds; open-adapted (open), clutter-adapted (clutter), edge-adapted with low-frequency echolocation calls (edge-low), and edge-adapted with medium- and high-frequency calls (edge-medium and edge-high). Clutter-adapted and edge-high species have been combined in this figure. Luck et al. 2013, figure 3. Reproduced with permission of Springer Science + Business Media. (b) A long-eared bat *Nyctophilus gouldi*. Photograph by Tracy Lee.

principle predicts that two species with identical fundamental niches cannot coexist indefinitely, as one species (the superior competitor) must eventually exclude the other (the inferior competitor; Gause 1934; Hardin 1960). Thus, fitness differences between (otherwise similar) species act to promote competitive exclusion (Chesson 2000).

The competitive exclusion principle can be extended to consider species with very similar niches (given the difficulty of determining the equivalence or otherwise of two niches in all possible dimensions of environmental space), and the situation where more than two such species occur together in space and time. Interestingly, this type of competitive exclusion has rarely been observed in natural environments, and the coexistence of multiple species with apparently identical niches is considered something of a paradox in ecology (Hutchinson 1961; Bode et al. 2011). Temporal and/or spatial variation in the availability of habitat, environmental stochasticity, mediating effects of predators and parasites, and differences between species in the distance or timing of dispersal may all contribute to the coexistence of multiple species with similar ecological niches (Shmida and Ellner 1984; Bode at al. 2011; Jiang et al. 2013). A framework developed by Chesson (2000) divides mechanisms that promote stable (ongoing) coexistence into two groups: stabilizing mechanisms such as niche differences (also known as resource partitioning), frequency-dependent predation, and mechanisms that rely on environmental variation in space and time; and equalizing mechanisms, which act to reduce fitness differences between species. As substantial fitness differences between species lead to competitive exclusion, they work in opposition to the stabilizing mechanisms of species coexistence.

Competitive superiority in urban environmental conditions is often proposed as the reason why certain species thrive in cities while others decline or disappear (Shochat et al. 2004, 2010; McKinney 2006). However, coexistence theory suggests that the small-scale patchiness (i.e., environmental heterogeneity across small spatial scales) and high rates of physical disturbance that lead to temporal variability in urban habitats are likely to assist multiple, similar species to coexist. Assuming that coexistence theory applies in urban environments, dominant urban species may have a considerable fitness advantage over the species that they replace, working in opposition to the stabilizing mechanisms arising from the variability of urban environmental conditions. As outlined earlier, Shochat et al. (2004, 2010) argue that urban bird communities are shaped by an increase in the availability of food and a decrease in predation pressure (Figure 4.4). Certain bird species (such as the Inca dove *Scardafella inca* and mourning dove *Zenaida macroura* in Phoenix, Arizona) are highly efficient foragers in urban habitats – where predation risk is lower than in their original, desert environment – to the point where there is little food left over for other granivorous species. This gives the Inca dove and mourning dove a fitness advantage over other native granivores in urban habitats. The lower abundance of certain predators in cities may also reduce the mediating effect of predation on highly abundant species, another stabilizing mechanism that can promote coexistence. However, competition for the same resources should be distinguished from the effects of bio-physical changes in urban environments that favour species with certain fundamental niches over others (e.g., Marzluff 2005).

4.3 Ecological drift: modelling stochasticity in urban communities

Neutral models in ecology emphasize ecological drift, which results from randomness or stochasticity. The importance of ecological drift in structuring ecological communities has been a topic of some debate in ecology. Hubbell (2001) proposed the unified neutral theory of biodiversity and biogeography to explain patterns of species richness and relative abundance within a given trophic level of a community, where multiple species share a similar niche and compete for similar resources. If these species are considered to be ecologically equivalent (i.e., each individual of each species is equally likely to reproduce, die, migrate, or evolve into a new species), then randomness can have a substantial influence on community structure. A key principle of the unified neutral theory is that community dynamics is a zero-sum game – no species in a community can increase in abundance without a matching decrease in the collective abundance of the other species (Hubbell 2001).

In this theory, Hubbell (2001) uses the term neutral to describe the assumption of per-capita ecological equivalence, and defines ecological drift as essentially identical to demographic stochasticity (chance births and deaths within a population – see Chapters 2 and 3). While ecologists had previously recognized that this kind of randomness could influence both the abundance and coexistence of species, and thus the composition of communities (e.g., Chesson and Warner 1981; Adler et al. 2007), Hubbell's neutral theory increased the appreciation of ecological drift as an important process in community ecology (Vellend 2010). Although it is unlikely that drift is the only process operating in most ecological communities, it may still be an important process, particularly where there is little functional difference between species (i.e., where selection is weak; Vellend 2010). Ecological drift can also interact with dispersal and/or evolutionary diversification to structure communities. For example, the relative abundance of tree species in 50-ha, permanent plots in the tropical rainforests of Pasoh Forest Reserve, Malaysia and Barro Colorado Island, Panama is well described by a neutral model that includes ecological drift and dispersal (Hubbell 2001).

But how important is ecological drift in shaping communities in urban environments? Obviously, this depends in part on the importance of drift relative to that of selection, dispersal and diversification in a given community prior to urbanization.

But it also depends on how the processes of urbanization alter the balance between ecological drift and the other three community-level processes. Habitat fragmentation and isolation by urban infrastructure, such as roads, may limit dispersal for many terrestrial animals, as well as for plants without wind-dispersed seeds (Trombulak and Frissell 2000; Develey and Stouffer 2001; Desender et al. 2005; also see Chapter 3). This may increase the importance of ecological drift in shaping these relatively isolated urban communities. The importance of ecological drift may also increase as the total number of individuals in a community declines (e.g., after a disturbance event; Vellend 2010). This is analogous to the case in population genetics where accelerated genetic drift follows a population decline or bottleneck (Ellstrand and Elam 1993). Certain types of urban ecological communities are likely to experience more intense and/or more frequent disturbance than their non-urban counterparts (e.g., Ishitani et al. 2003). This may in turn amplify the significance of chance immigration or local extinction of different species within the community in shaping future composition and relative abundance. On the other hand, processes of urbanization such as changes to climate, hydrological regimes or noise and light regimes may increase selection for certain species that can tolerate or thrive in the new conditions (e.g., Raghu et al. 2000; Bonier et al. 2007), and strong selection is likely to override the effects of ecological drift (Vellend 2010).

A study of spider, bee and bird communities across three cities in Switzerland found that the composition of communities was influenced more strongly by environmental variables, such as aspect, solar exposure, and the proportion of woody vegetation in a particular radius around the study site, than by spatial variables (Sattler et al. 2010). This suggested a stronger role of selection than of dispersal in shaping the composition and relative abundance of species within these communities. However, the total variation in community structure explained by the analysis was quite modest, ranging from 4% for bee communities in Lucerne and Lugano, to 30% for the bird community in Lugano, with an average explained variance across the nine communities of 12.5%. This means that between 70 and 96% of the variation in the composition of these communities was not explained by the environmental or spatial variables considered. It is possible that some of the unexplained variation in the composition of these communities resulted from ecological drift following repeated disturbance of the urban environment by humans (Sattler et al. 2010), which reduced the influence of selection and dispersal on community structure. Another explanation is that the analysis did not include some environmental variables that were important drivers of the composition of each community. Sattler et al. (2010) proposed that the near-absence of spatial structure (and therefore dispersal) in the communities they studied was typical of many urban ecological communities, in contrast to other studies of community composition in urban habitats which have found an important spatial component in community structure (Parris 2006; Chytrý et al. 2012). There is a clear need for further empirical and theoretical research on the importance of drift in shaping urban ecological communities.

4.4 Dispersal: the movement of individuals through space

Dispersal is the movement of individuals from one place to another, as propagules (seeds and spores), larvae, juveniles or adults. Dispersal can play a key role in structuring ecological communities, as individuals of different species arrive (immigrate) and/or leave (emigrate) over time. The influence of dispersal on community dynamics depends on the size, location and composition of the source communities (from where the dispersing individuals come) and receiver communities (to where they disperse), as well as the connectedness or isolation of each community (Leibold et al. 2004; Vellend 2010). In turn, connectedness or isolation depends on both the geographic distance between communities and the probability that individuals will travel successfully through the space that separates them (Leibold et al. 2004; Moritz et al. 2013). In natural environments, the composition of a local community is strongly influenced by that of its regional community, as the species in the former are necessarily a subset of the pool of species in the latter (Figure 4.8). Similarly, the composition of a regional community is strongly influenced by that of the continental or global community (Vellend 2010). However, in

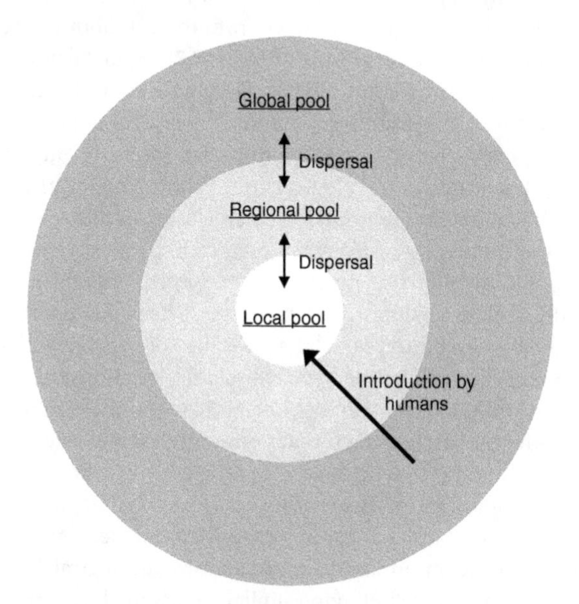

Figure 4.8 Schematic diagram of the composition of local, regional and global ecological communities. Through the dispersal of species, the composition of the local community is linked to that of the regional community, and the composition of the regional community is linked to that of the global community. However, in urban environments, the local species pool is also linked to the global species pool via the introduction by humans of species that are not present in the regional community.

urban environments, humans can introduce species into a local community that are not present in its regional community, thereby increasing the effective size of its source pool – potentially quite dramatically (Figure 4.8). This reduced dependence on the regional species pool contributes to the high species richness of many urban plant communities, for example (Hope et al. 2003; Knapp et al. 2008).

Community ecologists have developed a range of theories and models to explain how dispersal affects the coexistence of species and other aspects of community composition, including species richness and relative abundance. Some of these theories focus exclusively on dispersal, but most consider dispersal in conjunction with one or more of the other community-level processes (selection, ecological drift and/or evolutionary diversification). For example, the equilibrium theory of island biogeography considers drift and dispersal (MacArthur and Wilson 1963, 1967), the unified neutral theory of biodiversity and biogeography considers ecological drift, speciation and dispersal (Hubbell 2001), and metacommunity theory considers selection, ecological drift and dispersal (Hanski and Gilpin 1991; Leibold et al. 2004). The differential dispersal model of Bode et al. (2011) focuses on dispersal, and predicts that inter-specific differences in dispersal ability alone will allow otherwise identical species to coexist, if the habitat patches they occupy are located at irregular distances from each other.

Metacommunity theory is an extension of metapopulation theory (Levins 1969; see Section 3.2.1), which in turn developed from the equilibrium theory of island biogeography (MacArthur and Wilson 1963, 1967). A metacommunity can be defined as a group of local communities occupying a set of habitat patches that are linked by the dispersal of multiple, potentially interacting species (Leibold et al. 2004). Different perspectives on metacommunities variously emphasize the processes of dispersal, selection and ecological drift. Where a metacommunity consists of permanent habitat patches surrounded by an altered or inhospitable landscape, the number of species occupying a patch is predicted to increase with patch area as the probability of local extinction decreases (the species–area relationship; Hanski 1994). Secondly, isolated habitat patches are predicted to support fewer species per unit area than well-connected patches, because of a lower probability of colonization from another patch (the species–isolation relationship; Hanski 1994). If environmental conditions vary across habitat patches, these may favour different species in different patches, such that both dispersal and selection affect local community composition (the species-sorting perspective; Leibold et al. 2004). Dispersal remains important in this case because it allows changes in the composition of local communities to track changes in local environmental conditions. The neutral perspective of metacommunity theory emphasizes ecological drift and random dispersal events, with all species having similar fitness in all patches.

A study in Melbourne, Australia, assessed the applicability of metacommunity theory to frog communities at 104 ponds across an urban–rural gradient (Parris 2006). Poisson regression modelling identified an increase in species richness with patch area (pond size) and a decrease in species richness with increasing

Figure 4.9 The southern brown tree frog *Litoria ewingii* persists in ponds with a vertical wall in Melbourne, Australia. Photograph by Kirsten M. Parris.

patch isolation, as measured by surrounding road cover. Holding all other variables constant, species richness was predicted to be 2.8–5.5 times higher at the largest pond than at the smallest, while the species richness at the most-isolated pond was predicted to be 5.3–8.3 times lower than at the least-isolated pond. Environmental conditions were also important; the species richness at ponds with a surrounding vertical wall was predicted to be about half that at ponds with a gently sloping edge. Vertical walls make a pond unsuitable for breeding for many ground-dwelling frogs in the study area. These frogs are not able to climb vertical surfaces; if they did manage to breed in a walled pond, their offspring would be trapped in the pond and drowned following metamorphosis from tadpole to juvenile frog (Parris 2006). The southern brown tree frog *Litoria ewingii* (Figure 4.9) can climb vertical surfaces and was the only species regularly encountered at ponds with a vertical wall. These results demonstrate the applicability of metacommunity theory to this kind of urban system, as well as the importance of dispersal and selection in structuring communities of pond-dwelling frogs. Further evidence that dispersal and selection influence the composition of ecological communities at urban wetlands comes from a study of amphibians, reptiles and invertebrates at 201 wetlands in Colorado, USA (Johnson et al. 2013). All these taxonomic groups demonstrated lower species richness at wetlands in urban areas than those in agricultural or grassland areas, with both local environmental variables and the isolation of ponds by roads contributing to this pattern.

4.5 Diversification: the evolution of new lineages in urban environments

Diversification is the evolution of new lineages (including new genotypes, forms, varieties, sub-species and species) from existing lineages. In micro-organisms such as bacteria and archaea, diversification commonly arises via mutation or horizontal gene transfer (Nemergut et al. 2013). Compared to larger organisms, microbes can evolve very rapidly, particularly in circumstances where selection pressure is strong (Rainey and Travisano 1998; Rensing et al. 2002; Nemergut et al. 2013). For example, urban wastewater-treatment plants can act as hot spots for the evolution of antibiotic resistance in bacteria (Auerbach et al. 2007; Novo et al. 2013; Rizzo et al. 2013). The combination of antibiotic residues, antibiotic-resistant bacteria and non-resistant bacteria favours selection for and the horizontal transfer of antibiotic-resistance genes between individuals in wastewater (Novo et al. 2013). For example, antibiotic resistance in treated wastewater was positively correlated with the occurrence of tetracycline residues in raw wastewater and high temperatures at a treatment in Portugal; the rate of resistance to tetracycline and amoxicillin in treated wastewater varied from 1 to 60% among different groups of bacteria (Novo et al. 2013). An earlier study at the same plant found that wastewater treatment led to a three-fold increase in the prevalence of ciprofloxacin-resistant enterococci, probably due to a combination of selection for *Enterococcus faecium,* which has a high prevalence of resistance to ciprofloxacin, and horizontal gene transfer (da Silva et al. 2006).

In macro-organisms, a range of processes can lead to evolutionary diversification, many of which involve a level of reproductive isolation (Coyne and Orr 2004; Etienne et al. 2007). For example, allopatric diversification occurs when populations of a sexually-reproducing species become geographically isolated and gene flow between them is substantially reduced. Over time, genetic differences between the populations increase through genetic drift and/or selection (i.e., adaptation to local conditions), sometimes to the point where they become reproductively incompatible and are therefore considered as separate species. If one of the isolated populations is very small, a combination of founder effects and faster genetic drift may lead to faster speciation. This is known as peripatric speciation, and most commonly occurs when only a few individuals of a species colonize an isolated location such as an island (Barton and Charlesworth 1984; Seddon and Tobias 2007). Populations of a species may also become genetically isolated despite sharing the same geographic area (sympatric speciation); this can occur through hybridization and/or polyploidy (Rieseberg et al. 2006; Mallet 2007; Abbott et al. 2013). In the latter case, errors during cell division lead to certain individuals in a population carrying multiple sets of chromosomes, and these polyploids are usually unable to breed successfully with diploid individuals of the parent species. For example, a combination of human action, hybridization and polyploidy has led to the evolution of at least three new plant species in urban environments in the UK since 1700 (Thomas 2015).

New lineages can also arise through the process of sexual selection (Panhuis et al. 2001). Sexual selection results from the differential mating success of individuals within a population, and is often driven by a combination of male–male competition for mates and female preference for certain male traits. Sexual selection can lead to speciation when changes in both mate preference and one or more secondary sexual traits result in the reproductive isolation of populations within a species (Panhuis et al. 2001). Secondary sexual traits are features that distinguish the sexes, but are not necessary for reproduction. Morphological examples include the bright coloration of many male birds, the mane of a male lion and the antlers of a male deer, while behavioural examples include the song of male birds and the advertisement call of male frogs. Secondary sexual traits may be correlated with one or more aspects of male quality (e.g., energy reserves, body condition, body size), and therefore act as an honest signal of quality to both the females with whom they hope to mate and the males with whom they are competing (Hoelzer 1989; Moller and Pomiankowski 1993).

Urbanization could promote allopatric diversification, including speciation, if two conditions were met – an urban population of a species became isolated from its rural counterparts, and genetic drift and/or selection within the urban population resulted in adaptation to local conditions and thus genetic differentiation between the urban and rural populations. However, there are very few confirmed examples of allopatric speciation following urbanization. One example of allopatric diversification in progress is that of the house finch *Carpodacus mexicanus* in Arizona, USA. Individuals belonging to an urban population of this species in the city of Tucson have a longer, deeper and stronger bill than those from a nearby rural population in the Saguaro Desert (Badyaev et al. 2008). House finches in urban areas often forage at bird-feeders stocked with sunflower seeds, which are larger and harder than the seeds that form much of their diet in desert areas. Introduction of this novel food source appears to have exerted strong selection pressure on bill development and morphology in urban house finches, leading to a divergence of bill size and bite force between urban and rural populations. These adaptations to the urban food source are associated with genetic divergence between the populations, as well as differences in the structure of the male's courtship song – the urban birds with longer, deeper bills sing songs with slower trill rates (fewer notes per second) and a wider frequency range than the desert birds (Badyaev et al. 2008).

Evolutionary diversification in cities may also occur via sexual selection, if characteristics of the urban environment cause a change in mate preference and an associated change in a key secondary sexual trait. In the house-finch example described above, morphological adaptation to the urban food source (a longer, deeper bill) was correlated with changes in male courtship song, a secondary sexual trait that plays an important role in female mate choice. This difference in song may be acting to reinforce the local morphological adaptation, and to reduce gene flow between urban and rural populations of house finches (Badyaev et al. 2008). Many other species of birds are known to sing differently in urban than in rural habitats, including singing at a higher pitch to reduce acoustic interference from low-frequency urban noise (a frequency shift), singing more loudly (an amplitude shift), singing shorter songs, and singing more slowly (fewer syllables per second and fewer syllables per song; Slabbekoorn and Peet 2003; Brumm 2004; Nemeth and Brumm 2009; Potvin et al. 2011; Potvin and Parris 2012; Parris and McCarthy 2013; also see Section 3.2.5). However, no other study to date has found these characteristics of urban birdsong to be correlated with either a morphological or genetic difference between urban and rural populations (e.g., Partecke et al. 2004; Leader et al. 2008; Potvin et al. 2013).

Instead, song differences between urban and rural birds could arise through vocal plasticity (the ability to change the characteristics of a song to suit conditions such as the level and frequency of ambient noise) and/or cultural evolution (the non-random transmission of song types across generations; Halfwerk and Slabbekoorn 2009; Luther and Baptista 2010). Like birdsong, the advertisement call of male frogs plays an important role in mate choice and the reinforcement of inter-specific boundaries (Blair 1964; Wells 1977). In a number of frog species, males are known to call at a higher frequency (pitch) in urban noise and road-traffic noise (Parris et al. 2009; Hoskin and Goosem 2010; Cunnington and Fahrig 2010). At this stage, it is unclear whether these frequency shifts are the result of vocal plasticity or an early indication of sexual selection for higher-pitched calls.

4.6 Summary

This chapter considers the impact of urbanization on ecological communities, focusing on the four community-level processes of selection, ecological drift, dispersal and evolutionary diversification. While ecologists have developed a wide assortment of hypotheses, theories and models to explain the structure of ecological communities, few of these have been developed with urban environments in mind. However, this chapter demonstrates that many existing ecological theories and models – including niche theory, environmental gradients, resource-competition models, food webs, trophic cascades and metacommunity theory – do apply in urban environments, and can be used to improve our understanding of the impacts of urbanization at the level of the ecological community. It is likely that other community theories also apply in these environments, but are yet to be examined. Urbanization changes ecological communities by changing environmental conditions in cities, which in turn favours certain species over others (selection). The resulting changes in the identity and relative abundance of species within a community alter the interactions between species at different trophic levels, including autotrophs, herbivores and predators, leading to further changes in relative abundance and/or local extinction of some species. The fragmentation and isolation of habitats by roads and other urban infrastructure may constrain the dispersal of individuals from one habitat patch to another, with important consequences for a range of taxonomic groups. Reduced dispersal may increase the influence of random processes, such as demographic stochasticity, on community structure. There is also evidence of evolutionary diversification in urban habitats, where urban populations of a species become adapted to local conditions and/or reproductively isolated from rural populations. In the next chapter, I consider the next level of ecological organisation, examining the responses of entire ecosystems to urbanization.

Study questions

1 Describe the four community-level processes discussed in this chapter. Do you think that some of these are more important in urban environments than others?
2 Give three examples of the application of niche theory to the study of ecological communities in cities.
3 Draw an urban food web for a town or city you know.
4 How does urbanization affect dispersal? How might changes to dispersal across the urban landscape influence the structure of ecological communities?
5 Do the processes of urbanization increase the importance of ecological drift?
6 How might urbanization promote evolutionary diversification, including speciation, in cities?
7 With examples, describe three community-ecological theories that apply in urban environments.

References

Abbott, R., Albach, D., Ansell, S. *et al.* (2013) Hybridization and speciation. *Journal of Evolutionary Biology*, **26**, 229–246.

Adler, P.B., HilleRisLambers, J., and Levine, J.M. (2007) A niche for neutrality. *Ecology Letters*, **10**, 95–104.

Alberti, M. (2008) *Advances in Urban Ecology: Integrating Human and Ecological Processes in Urban Ecosystems*, Springer, New York.

Armstrong, A.J. (2009) Distribution and conservation of the coastal population of the black-headed dwarf chameleon *Bradypodion melanocephalum* in KwaZulu-Natal. *African Journal of Herpetology*, **58**, 85–97.

Auerbach, E.A., Seyfried, E.E., and McMahon, K.D. (2007) Tetracycline resistance genes in activated sludge wastewater treatment plants. *Water Research*, **41**, 1143–1151.

Austin, M.P. (1987) Models for the analysis of species response to environmental gradients. *Vegetatio*, **69**, 35–45.

Badyaev, A.V., Young, R.L., Oh, K.P. *et al.* (2008) Evolution on a local scale: developmental, functional, and genetic bases of divergence in bill form and associated changes in song structure between adjacent habitats. *Evolution*, **62**, 1951–1964.

Barton, N.H. and Charlesworth, B. (1984) Genetic revolutions, founder effects, and speciation. *Annual Review of Ecology and Systematics*, **15**, 133–164.

Begossi, A. (1996) Use of ecological methods in ethnobotany: diversity indices. *Economic Botany*, **50**, 280–289.

Bennett, A.F., Lumsden, L.F., Alexander, J.S.A. *et al.* (1991) Habitat use by arboreal mammals along an environmental gradient in north-eastern Victoria. *Wildlife Research*, **18**, 125–146.

Blair, R.B. (1996) Birds and butterflies along an urban gradient: surrogate taxa for assessing biodiversity? *Ecological Applications*, **9**, 164–170.

Blair, R.B. (1999) Land use and avian species diversity along an urban gradient. *Ecological Applications*, **6**, 506–519.

Blair, R.B. and Launer, A.E. (1997) Butterfly diversity and human land use: species assemblages along an urban gradient. *Biological Conservation*, **80**, 113–125.

Blair, W.F. (1964) Isolating mechanisms and interspecies interactions in anuran amphibians. *Quarterly Review of Biology*, **39**, 333–344.

Bode, M., Bode, L., and Armsworth, P.R. (2011) Different dispersal abilities allow reef fish to coexist. *Proceedings of the National Academy of Sciences of the United States of America*, **108**, 16317–16321.

Bolnick, D.I., Amarasekare, P., Araujo, M.S. *et al.* (2011) Why intraspecific trait variation matters in community ecology. *Trends in Ecology and Evolution*, **26**, 183–192.

Bonier, F., Martin, P.R., and Wingfield, J.C. (2007) Urban birds have broader environmental tolerance. *Biology Letters*, **3**, 670–673.

Borcard, D., Legendre, P., and Drapeau, P. (1992) Partialling out the spatial component of ecological variation. *Ecology*, **73**, 1045.

Bridgman, H., Warner, R., and Dodson, J. (1995) *Urban Biophysical Environments*, Oxford University Press, Melbourne.

Brumm, H. (2004) The impact of environmental noise on song amplitude in a territorial bird. *Journal of Animal Ecology*, **73**, 434–440.

Buckley, L.B. and Jetz, W. (2010) Lizard community structure along environmental gradients. *Journal of Animal Ecology*, **79**, 358–365.

Burgman, M.A., Breininger, D.R., Duncan, B.W. *et al.* (2001) Setting reliability bounds on habitat suitability indices. *Ecological Applications*, **11**, 70–78.

Burke, C., Steinberg, P., Rusche, D. *et al.* (2011) Bacterial community assembly based on functional genes rather than species. *Proceedings of the National Academy of Sciences*, **108**, 14288–14293.

Busby, J.R. (1991) A bioclimate analysis and prediction system, in *Nature Conservation: Cost Effective Biological Surveys and Data Analysis* (eds C.R. Margules and M.P. Austin), CSIRO, Canberra, pp. 64–68.

Carpenter, G., Gillison, A.N., and Winter, J. (1993) A flexible modelling procedure for mapping potential distributions of plants and animals. *Biodiversity and Conservation*, **2**, 667–680.

Chace, J.F. and Walsh, J.J. (2006) Urban effects on native avifauna: a review. *Landscape and Urban Planning*, **74**, 46–69.

Chase, J.M., Liebold, M.A., Downing, A.L. *et al.* (2000) The effects of productivity, herbivory, and plant species turnover in grassland food webs. *Ecology*, **81**, 2485–2497.

Chesson, P.L. (2000) Mechanisms of maintenance of species diversity. *Annual Review of Ecology and Systematics*, **31**, 343–366.

Chesson, P.L. and Warner, R.R. (1981) Environmental variability promotes coexistence in lottery competitive systems. *The American Naturalist*, **117**, 923–943.

Chytry, M., Lososova, Z., Horsak, M. *et al.* (2012) Dispersal limitation is stronger in communities of microorganisms across Central European cities. *Journal of Biogeography*, **39**, 1101–1111.

Clements, F.E. (1916) *Plant Succession: An Analysis of the Development of Vegetation*, Carnegie Institute of Washington, Washington DC.

Clements, F.E. (1936) Nature and structure of the climax. *Journal of Ecology*, **24**, 252–284.

Cottenie, K. (2005) Integrating environmental and spatial processes in ecological community dynamics. *Ecology Letters*, **8**, 1175–1182.

Coyne, J.A. and Orr, H.A. (2004) *Speciation*, Sinauer Associates, Sunderland, MA.

Cunnington, G.M. and Fahrig, L. (2010) Plasticity in the vocalizations of anurans in response to traffic. *Acta Oecologica*, **36**, 463–470.

da Silva, M.F., Tiago, I., Verissimo, A. *et al.* (2006) Antibiotic resistance of enterococci and related bacteria in an urban wastewater treatment plant. *FEMS Microbiology Ecology*, **55**, 322–329.

Desender, K., Small, E., Gaublomme, E. *et al.* (2005) Rural-urban gradients and the population genetic structure of woodland ground beetles. *Conservation Genetics*, **6**, 51–62.

Develey, P.F. and Stouffer, P.C. (2001) Effects of roads on movements by understory birds in mixed-species flocks in Central Amazonian Brazil. *Conservation Biology*, **15**, 1416–1422.

Duncan, R.P., Clemants, S.E., Corlett, R.T. *et al.* (2011) Plant traits and extinction in urban areas: a meta-analysis of 11 cities. *Global Ecology and Biogeography*, **20**, 509–519.

Elith, J., Graham, C.H., Anderson, R.P. *et al.* (2006) Novel methods improve prediction of species distributions from occurrence data. *Ecography*, **29**, 129–151.

Elith, J. and Leathwick, J.R. (2009) Species distribution models: ecological explanation and prediction across space and time. *Annual Review of Ecology, Evolution, and Systematics*, **40**, 677–697.

Ellstrand, N.C. and Elam, D.R. (1993) Population genetic consequences of small population size: implications for plant conservation. *Annual Review of Ecology and Systematics*, **24**, 217–242.

Etienne, R.S., Emile, M., Apol, F. *et al.* (2007) Modes of speciation and the neutral theory of biodiversity. *Oikos*, **116**, 241–258.

Faeth, S.H., Warren, P.S., Shochat, E. *et al.* (2005) Trophic dynamics in urban communities. *BioScience*, **55**, 399–407.

Ferrier, S., Drielsma, M., Manion, G. *et al.* (2002) Extended statistical approached to modelling spatial pattern in biodiversity in northeast New South Wales. II. Community-level modelling. *Biodiversity and Conservation*, **11**, 2309–2338.

Ferrier, S., Watson, G., Pearce, J. *et al.* (2002) Extended statistical approached to modelling spatial patterns in biodiversity in northeast New South Wales. I. Species-level modelling. *Biodiversity and Conservation*, **11**, 2275–2307.

Gates, K.K., Vaughn, C.C., and Julian, J.P. (2015) Developing environmental flow recommendations for freshwater mussels using the biological traits of species guilds. *Freshwater Biology*, **60**, 620–635.

Gause, G.F. (1934) *The Struggle for Existence*, Williams & Wilkins, Baltimore.

Gerhardt, H.C. and F. Huber (eds). (2002) *Acoustic Communication in Insects and Anurans: Common Problems and Diverse Solutions*. University of Chicago Press, Chicago.

Gleason, H.A. (1917) The structure and development of the plant association. *Bulletin of the Torrey Botanical Club*, **44**, 463–481.

Gleason, H.A. (1926) The individualistic concept of the plant association. *Bulletin of the Torrey Botanical Club*, **53**, 1–20.

Grinnell, J. (1917) The niche-relationships of the California Thrasher. *The Auk*, **34**, 427–433.

Hahs, A.K. and McDonnell, M.J. (2006) Selecting independent measures to quantify Melbourne's urban-rural gradient. *Landscape and Urban Planning*, **78**, 435–448.

Halfwerk, W. and Slabbekoorn, H. (2009) A behavioural mechanism explaining noise-dependent frequency use in urban birdsong. *Animal Behaviour*, **78**, 1301–1307.

Hamer, A.J. and Parris, K.M. (2013) Predation modifies larval amphibian communities in urban wetlands. *Wetlands*, **33**, 641–652.

Hamer, A.J. and Parris, K.M. (2011) Local and landscape determinants of amphibian communities in urban ponds. *Ecological Applications*, **21**, 378–390.

Hanski, I. (1994) A practical model of metapopulation dynamics. *Journal of Animal Ecology*, **63**, 151–162.

Hanski, I. and Gilpin, M. (1991) Metapopulation dynamics: brief history and conceptual domain, in *Metapopulation Dynamics* (eds M.E. Gilpin and I. Hanski), Academic Press, London, pp. 3–16.

Hardin, G. (1960) The competitive exclusion principle. *Science*, **131**, 1292–1297.

Hatcher, M.J., Dick, J.T.A., and Dunn, A.M. (2006) How parasites affect interactions between competitors and predators. *Ecology Letters*, **9**, 1253–1271.

Hironaka, Y. and Koike, F. (2013) Guild structure in the food web of grassland arthropod communities along an urban-rural landscape gradient. *Ecoscience*, **20**, 148–160.

Hoelzer, G.A. (1989) The good parent process of sexual selection. *Animal Behaviour*, **38**, 1067–1078.

Hope, D., Gries, C., Zhu, W. *et al.* (2003) Socioeconomics drive urban plant diversity. *Proceedings of the National Academy of Sciences of the United States of America*, **100**, 8788–8792.

Hoskin, C.J. and Goosem, M.W. (2010) Road impacts on abundance, call traits, and body size of rainforest frogs in northeast Australia. *Ecology and Society* **15**, 15. http://www.ecologyandsociety.org/vol15/iss3/art15

Hoverman, J.T., Davis, C.J., Werner, E.E. *et al.* (2011) Environmental gradients and the structure of freshwater snail communities. *Ecography*, **34**, 1049–1058.

Howard, L. (1833) *Climate of London Deduced from Meteorological Observations*, Harvey and Darton, London.

Hu, L., Li, M., and Li, Z. (2010) Geographical and environmental gradients of lianas and vines in China. *Global Ecology and Biogeography*, **19**, 554–561.

Hubbell, S.P. (2001) *The Unified Neutral Theory of Biodiversity and Biogeography*, Princeton University Press, Princeton.

Huijbers, C.M., Schlacher, T.A., Schoeman, D.S. *et al.* (2015) Limited functional redundancy in vertebrate scavenger guilds fails to compensate for the loss of raptors from urbanized sandy beaches. *Diversity and Distribution*, **21**, 55–63.

Hutchinson, G.E. (1957) Concluding remarks. *Cold Spring Harbor Symposium on Quantitative Biology*, **22**, 415–427.

Hutchinson, G.E. (1961) The paradox of the plankton. *The American Naturalist*, **95**, 137–145.

Ishitani, M., Kotze, J., and Niemela, J. (2003) Changes in carabid beetle assemblages across an urban-rural gradient in Japan. *Ecography*, **26**, 481–489.

Jacob, B.G., Arheart, K.L., Griffith, D.A. *et al.* (2005) Evaluation of environmental data for identification of *Anopheles* (Diptera: Culicidae) aquatic larval habitats in Kisumu and Malindi, Kenya. *Entomology Society of America*, **42**, 751–755.

Jiang, T., Lu, G., Sun, K. *et al.* (2013) Coexistence of *Rhinolophus affinis* and *Rhinolophus pearsoni* revisited. *Acta Theriologica*, **58**, 47–53.

Johnson, P.T.J., Hoverman, J.T., McKenzie, V.J. *et al.* (2013) Urbanization and wetland communities: applying metacommunity theory to understanding the local and landscape effects. *Journal of Applied Ecology*, **50**, 34–42.

Knapp, S., Kuhn, I., Schweiger, O. *et al.* (2008) Challenging urban species diversity: contrasting phylogenetic patterns across plant functional groups in Germany. *Ecological Letters*, **11**, 1054–1064.

Konstantinidis, K.T., Ramette, A., and Tiedje, J.M. (2006) The bacterial species definition in the genomic era. *Philosophical Transactions of the Royal Society B*, **361**, 1929–1940.

Lawson, D.M., Lamar, C.K., and Schwartz, M.W. (2008) Quantifying plant population persistence in human-dominated landscapes. *Conservation Biology*, **22**, 922–928.

Lawton, J. (1999) Are there general laws in ecology? *Oikos*, **84**, 177.

Leader, N., Geffen, E., Mokady, O. *et al.* (2008) Song dialects do not restrict gene flow in an urban population of the orange-tufted sunbird, *Nectarinia osea*. *Behavioral Ecology and Sociobiology*, **62**, 1299–1305.

Leibold, M.A., Holyoak, M., Mouquet, N. *et al.* (2004) The metacommunity concept: a framework for multi-scale community ecology. *Ecology Letters*, **7**, 601–613.

Levins, R. (1969) Some demographic and genetic consequences of environmental heterogeneity for biological control. *Bulletin of the Entomological Society of America*, **15**, 237–240.

Lim, H.C. and Sodhi, N.S. (2004) Responses of avian guilds to urbanization in a tropical city. *Landscape and Urban Planning*, **66**, 199–215.

Liu, X., Zhang, W., Li, H. *et al.* (2013) Modeling patch characteristics of farmland loss for site assessment in urban fringe of Beijing, China. *Chinese Geographical Science*, **23**, 365–377.

Luck, G.W., Smallbone, L., Threlfall, C.G. *et al.* (2013) Patterns in bat functional guilds across multiple urban centres in south-eastern Australia. *Landscape Ecology*, **28**, 455–469.

Luck, M. and Wu, J. (2002) A gradient analysis of urban landscape pattern: a case study from the Phoenix metropolitan region, Arizona, USA. *Landscape Ecology*, **17**, 327–339.

Luther, D. and Baptista, L. (2010) Urban noise and the cultural evolution of bird songs. *Proceedings of the Royal Society B*, **277**, 469–473.

MacArthur, R.H. (1960) On the relative abundance of species. *The American Naturalist*, **94**, 25–36.

MacArthur, R.H. and Wilson, E.O. (1963) An equilibrium theory of insular zoogeography. *Evolution*, **17**, 373–387.

MacArthur, R.H. and Wilson, E.O. (1967) *The Theory of Island Biogeography*, Princeton University Press, Princeton.

Mallet, J. (2007) Hybrid speciation. *Nature*, **446**, 279–283.

Marzluff, J.M. (2005) Island biogeography for an urbanizing world: how extinction and colonization may determine biological diversity in human-dominated landscapes. *Urban Ecosystems*, **8**, 157–177.

Marzluff, J.M. (2001) Worldwide urbanization and its effects on birds, in *Avian Ecology and Conservation in an Urbanizing World* (eds J.M. Marzluff, R. Bowman, and R. Donnelly), Kluwer, Boston, pp. 19–38.

McDonnell, M.J. and Hahs, A.K. (2008) The use of gradient analysis studies in advancing our understanding of the ecology of urbanizing landscapes: current status and future directions. *Landscape Ecology*, **23**, 1143–1155.

McDonnell, M.J. and Pickett, S.T.A. (1990) Ecosystem structure and function along urban-rural gradients: an unexploited opportunity for ecology. *Ecology*, **71**, 1232–1237.

McDonnell, M.J., Picket, S.T.A., Groffman, P.M. *et al.* (1993) Ecosystem processes along an urban-to-rural gradient. *Urban Ecosystems*, **1**, 21–36.

McKinney, M.L. (2008) Effects of urbanization on species richness: a review of plants and animals. *Urban Ecosystems*, **11**, 161–176.

McKinney, M.L. (2006) Urbanization as a major cause of biotic homogenization. *Biological Conservation*, **127**, 247–260.

Moller, A.P. and Pomiankowski, A. (1993) Why have birds got multiple sexual ornaments? *Behavioral Ecology and Sociobiology*, **32**, 167–176.

Moritz, C., Meynard, C.N., Devictor, V. *et al.* (2013) Disentangling the role of connectivity, environmental filtering, and spatial structure on metacommunity dynamics. *Oikos*, **122**, 1–10.

Nemergut, D.R., Schmidt, S.K., Fukami, T. *et al.* (2013) Patterns and processes of microbial community assembly. *Microbiology and Molecular Biology Reviews*, **77**, 342–356.

Nemeth, E. and Brumm, H. (2009) Birds and anthropogenic noise: are urban songs adaptive? *The American Naturalist*, **176**, 465–475.

Niemela, J., Kotze, D.J., Venn, S. *et al.* (2002) Carabid beetle assemblages (Coleoptera, carabidae) across urban-rural gradients: an international comparison. *Landscape Ecology*, **17**, 387–401.

Novo, A., Andre, S., Viana, P. *et al.* (2013) Antibiotic resistance, antimicrobial residues and bacterial community composition in urban wastewater. *Water Research*, **47**, 1875–1887.

Pace, M.L., Cole, J.J., Carpenter, S.R. *et al.* (1999) Trophic cascades revealed in diverse ecosystems. *Trends in Ecology and Evolution*, **14**, 483–488.

Palmer, M.W. (1994) Variation in species richness: towards a unification of hypotheses. *Folia Geobotanica and Phytotaxonomica*, **29**, 511–530.

Panhuis, T.M., Butlin, R., Zuk, M. *et al.* (2001) Sexual selection and speciation. *Trends in Ecology and Evolution*, **16**, 364–371.

Parris, K.M. (2001) Distribution, habitat requirements, and conservation of the cascade treefrog (*Litoria pearsoniana*, Anura: Hylidae). *Biological Conservation*, **99**, 285–292.

Parris, K.M. (2006) Urban amphibian assemblages as metacommunities. *Journal of Animal Ecology*, **75**, 757–746.

Parris, K.M. and McCarthy, M.A. (2013) Predicting the effects of urban noise on the active space of avian vocal signals. *The American Naturalist*, **182**, 452–464.

Parris, K. M., Velik-Lord, M. and North, J.M.A. (2009) Frogs call at a higher pitch in traffic noise. *Ecology and Society* **14**, 25. http://www.ecologyandsociety.org/vol14/iss1/art25/

Partecke, J., Van't Hof, T., and Gwinner, E. (2004) Differences in the timing of reproduction between urban and forest European blackbirds (*Turdus merula*): result of phenotypic flexibility or genetic differences? *Proceedings of the Royal Society B*, **271**, 1995–2001.

Pauw, A., and Louw, K. (2012) Urbanization drives a reduction in functional diversity in a guild of nectar-feeding birds. *Ecology and Society* **17**, 27. http://www.ecologyandsociety.org/vol17/iss2/art27/

Peres-Neto, P.R. and Kembel, S.W. (2015) Phylogenetic gradient analysis: environmental drivers of phylogenetic variation across ecological communities. *Plant Ecology*, **216**, 709–724.

Phillips, S.J., Anderson, R.P., and Schapire, R.E. (2006) Maximum entropy modelling of species geographic distribution. *Ecological Modelling*, **190**, 231–259.

Pimm, S.L., Lawton, J.H., and Cohen, J.E. (1991) Food web patterns and their consequences. *Nature*, **350**, 669–674.

Polis, G.A., Myers, C.A., and Holt, R.D. (1989) The ecology and evolution of intraguild predation: potential competitors that eat each other. *Annual Review of Ecology and Systematics*, **20**, 297–330.

Potapova, M.G. and Charles, D.F. (2002) Benthic diatoms in USA rivers: distribution along spatial and environmental gradients. *Journal of Biogeography*, **29**, 167–187.

Potvin, D.A. and Parris, K.M. (2012) Song convergence in multiple urban populations of silvereyes (*Zosterops lateralis*). *Ecology and Evolution*, **2**, 1977–1984.

Potvin, D.A., Parris, K.M., and Mulder, R.A. (2011) Geographically pervasive effects of urban noise on frequency and syllable rate of songs and calls in silvereyes (*Zosterops lateralis*). *Proceedings of the Royal Society B*, **278**, 2464–2469.

Potvin, D.A., Parris, K.M., and Mulder, R.A. (2013) Limited genetic differentiation between acoustically divergent populations of urban and rural silvereyes (*Zosterops lateralis*). *Evolutionary Ecology*, **27**, 381–391.

Preston, F.W. (1948) The commonness, and rarity, of species. *Ecology*, **29**, 254–283.

Raghu, S., Clarke, A.R., Drew, R.A.I. *et al.* (2000) Impact of habitat modification on the distribution and abundance of fruit flies (*Diptera: Tephritidae*) in southeast Queensland. *Population Ecology*, **42**, 153–160.

Rainey, P.B. and Travisano, M. (1998) Adaptive radiation in a heterogeneous environment. *Nature*, **394**, 69–72.

Ramalho, C.E. and Hobbs, R.J. (2012) Time for a change: dynamic urban ecology. *Trends in Ecology and Evolution*, **27**, 179–188.

Rensing, C., Newby, D.T., and Pepper, I.L. (2002) The role of selective pressure and selfish DNA in horizontal gene transfer and soil microbial community adaptation. *Soil Biology and Biochemistry*, **34**, 285–296.

Rieseberg, L.H., Wood, T.E., and Baack, E.J. (2006) The nature of plant species. *Nature*, **440**, 524–527.

Ritchie, E.G. and Johnson, C.N. (2009) Predator interactions, mesopredator release, and biodiversity conservation. *Ecology Letters*, **12**, 982–998.

Rizzo, L., Manaia, C., Merlin, C. *et al.* (2013) Urban wastewater treatment plants as hotspots for antibiotic resistant bacteria and genes spread into the environment: A review. *Science of the Total Environment*, **447**, 345–360.

Root, R.B. (1967) The niche exploitation pattern of the blue-gray gnatcatcher. *Ecological Monographs*, **37**, 317–350.

Rosindell, J., Hubbell, S.P., and Etienne, R.S. (2011) The unified neutral theory of biodiversity and biogeography at age ten. *Trends in Ecology and Evolution*, **26**, 340–348.

Sadler, J.P., Small, E.C., Fiszpan, H. *et al.* (2006) Investigating environmental variation and landscape characteristics of an urban-rural gradient using woodland carabid assemblages. *Journal of Biogeography*, **33**, 1126–1138.

Saito, M. and Koike, F. (2009) The importance of past and present landscape for Japanese hares *Lepus brachyurus* along a rural-urban gradient. *Acta Theriologica*, **54**, 363–370.

Sattler, T., Borcard, D., Arlettaz, R. *et al.* (2010) Spider, bee, and bird communities in cities are shaped by environmental control and high stochasticity. *Ecology*, **91**, 3343–3353.

Seddon, N. and Tobias, J.A. (2007) Song divergence at the edge of Amazonia: an empirical test of the peripatric speciation model. *Biological Journal of the Linnean Society*, **90**, 173–188.

Shmida, A. and Ellner, S. (1984) Coexistence of plant species with similar niches. *Vegetatio*, **58**, 29–55.

Shochat, E. (2004) Credit or debit? Resource input changes population dynamics of city-slicker birds. *Oikos*, **106**, 622–626.

Shochat, E., Lerman, S.B., Anderies, J.M. *et al.* (2010) Invasion, competition, and biodiversity loss in urban ecosystems. *BioScience*, **60**, 199–208.

Shochat, E., Lerman, S., Katti, M. *et al.* (2004) Linking optimal foraging behavior to bird community structure in an urban-desert landscape: Field experiments with artificial food patches. *American Naturalist*, **164**, 232–243.

Simberloff, D. (2004) Community ecology: is it time to move on? *The American Naturalist*, **163**, 787–799.

Simberloff, D. and Dayan, T. (1991) The guild concept and the structure of ecological communities. *Annual Review of Ecology and Systematics*, **22**, 115–143.

Slabbekoorn, H. and Peet, M. (2003) Birds sing at a higher pitch in urban noise. *Nature*, **424**, 267.

Spellerberg, I.F. and Fedor, P.J. (2003) A tribute to Claude Shannon (1916-2001) and a plea for more rigorous use of species richness, species diversity, and the 'Shannon-Wiener' Index. *Global Ecology & Biogeography*, **12**, 177–179.

Stracey, C.M. and Robinson, S.K. (2012) Are urban habitats ecological traps for a native songbird? Season-long productivity, apparent survival, and site fidelity in urban and rural habitats. *Journal of Avian Biology*, **43**, 50–60.

Ter Braak, C.J.F. (1986) Canonical correspondence analysis: a new eigenvector technique for multivariate direct gradient analysis. *Ecology*, **67**, 1167–1179.

Ter Braak, C.J.F. (1987) The analysis of vegetation-environment relationship by canonical correspondence analysis. *Vegetatio*, **69**, 69–77.

Ter Braak, C.J.F. and Prentice, I.C. (1988) A theory of gradient analysis. *Advances in Ecological Research*, **18**, 271–317.

Terborgh, J. (1971) Distribution on environmental gradients: theory and a preliminary interpretation of distributional patterns in the avifauna of the Cordillera Vilcabamba, Peru. *Ecology*, **52**, 23–40.

Thomas, C.D. (2015) Rapid acceleration of plant speciation during the Anthropocence. *Trends in Ecology and Evolution*. 10.1016/j.tree.2015.05.009

Thompson, R.M., Brose, U., Dunne, J.A. *et al.* (2012) Food webs: reconciling the structure and function of biodiversity. *Trends in Ecology and Evolution*, **27**, 689–697.

Torok, S.J., Morris, C.J.G., Skinner, C. *et al.* (2001) Urban heat island features of southeast Australian towns. *Australian Meteorological Magazine*, **50**, 1–13.

Trombulak, S.C. and Frissell, C.A. (2000) Review of ecological effects of roads on terrestrial and aquatic communities. *Conservation Biology*, **14**, 18–30.

Van Rensburg, B.J., Peacock, D.S., and Robertson, M.P. (2009) Biotic homogenization and alien bird species along an urban gradient in South Africa. *Landscape and Urban Planning*, **92**, 233–241.

Vellend, M. (2010) Conceptual synthesis in community ecology. *The Quarterly Review of Biology*, **85**, 183–206.

Violle, C., Enquist, B.J., McGill, B.J. *et al.* (2012) The return of the variance: intraspecific variability in community ecology. *Trends in Ecology and Evolution*, **27**, 244–252.

Walsh, C.J., Fletcher, T.D., and Ladson, A.R. (2005) Stream restoration in urban catchments through redesigning stormwater systems: looking to the catchment to save the stream. *Journal of the North American Benthological Society*, **24**, 690–705.

Walsh, C.J., Waller, K.A., Gehling, J. *et al.* (2007) Riverine invertebrate assemblages are degraded more by catchment urbanization than by riparian deforestation. *Freshwater Biology*, **52**, 574–587.

Weaver, P.L. (2000) Environmental gradients affect forest structure in Puerto Rico's Luquilla Mountains. *Interciencia*, **25**, 254–259.

Wells, K.D. (1977) The social behaviour of anuran amphibians. *Animal Behaviour*, **25**, 666–693.

Whittaker, R.H. (1967) Gradient analysis of vegetation. *Biological Reviews*, **42**, 207–264.

Williams, N.S.G., McDonnell, M.J., and Seager, E.J. (2005) Factors influencing the loss of an endangered ecosystem in an urbanising landscape: a case study of native grasslands from Melbourne, Australia. *Landscape and Urban Planning*, **71**, 35–49.

Williams, S.E. and Hero, J.-M. (1998) Rainforest frogs of the Australian wet tropics: guild classification and the ecological similarity of declining species. *Proceedings of the Royal Society B*, **265**, 597–602.

Wintle, B.A., Elith, J., and Potts, J.M. (2005) Fauna habitat modelling and mapping: a review and case study in the Lower Hunter Central Coast region of NSW. *Austral Ecology*, **30**, 719–738.

CHAPTER 5

Ecosystem-level responses to urbanization

5.1 Introduction

Population ecology can be understood through the four population processes of births, deaths, immigration and emigration; its focus is on the individuals within a population. Similarly, community ecology can be understood through the four community-level processes of selection, dispersal, ecological drift and evolutionary diversification; its focus is on the species within a community. But how do we characterize the ecology of entire ecosystems? First, we need to clarify what we mean by the term ecosystem. A *system* is a group of parts that interact with each other via one or more processes, with emergent properties that arise from these interactions. As a consequence, the system as a whole is greater than the sum of its parts (Odum 1983). An *ecosystem* (short for ecological system) was defined by Tansley (1935) as the collection of living and non-living components in a given place or geographic area, and the network of interactions between them. Therefore, an ecosystem has four broad elements: the living element, comprised of all organisms from the smallest microbes to the tallest trees and the largest animals; the non-living element, comprised of air, water, light, heat and minerals (rock, soil, sand, dust, carbon and other nutrients); the interactions within and between the first two elements; and a designated, physical space within which all these elements exist (Tansley 1935; Keith et al. 2013). Biological, chemical and physical processes occur when two or more components of an ecosystem interact (Odum 1983); examples include photosynthesis, carbon sequestration, evapotranspiration, respiration, predation, nitrogen fixation, decomposition, precipitation, insolation, erosion, the weathering of rocks and minerals and the absorption of water and nutrients by soil (see Box 5.1).

Ecosystem ecology focuses on the ecosystem in its entirety, measuring and modelling the fluxes of energy and materials in, around and out of the system (Loreau 2010). Like ecological communities, ecosystems can be studied across a range of spatial scales from the local to the global, depending on the entity or question of interest (Tansley 1935; Odum 1983; Willis 1997; Loreau 2010). For example, let's consider the ecosystem formed by a watershed or hydrological

Ecology of Urban Environments, First Edition. Kirsten M. Parris.

Box 5.1 What are ecosystem processes, functions and services?

While the terms ecosystem process, ecosystem function and ecosystem service are widely used in ecology, environmental science and economics, their meanings and relationships to each other are often poorly defined. I provide some clear definitions here.

Ecosystem process: A process that occurs when two or more parts of an ecosystem interact (Odum 1983). Examples include insolation (sunlight falls on a rock or a tree), erosion (fast flowing water moves soil from one place to another), photosynthesis (a plant converts light energy from the sun into chemical energy), nitrogen fixation (bacteria within the root system of a legume convert atmospheric nitrogen into ammonia), herbivory (an animal eats a plant) and predation (an animal eats an animal). Important cycles within ecosystems, such as the carbon cycle and water cycle, are comprised of many different processes.

Ecosystem function: A function performed by an ecosystem – that is, something that an ecosystem *does*. Ecosystem functions rely on and arise from one or more ecosystem processes. Examples include primary production (which relies on precipitation, insolation, photosynthesis, and various processes that influence the concentration of bio-available nutrients in the soil), hydrological flows (which rely on precipitation, infiltration, evapotranspiration, percolation, etc.) and pollination (which relies on various processes that move pollen from the anthers of a flower to a receptive stigma, such as wind or the consecutive visitation of flowers of the same species by birds, insects and mammals). Some authors appear to use the terms ecosystem process and ecosystem function interchangeably (e.g., Costanza et al. 1997; Diaz & Cabido 1997; Crutsinger et al. 2006).

Ecosystem service: A benefit that humans derive, directly or indirectly, from ecosystem functions (Costanza et al. 1997; Millennium Ecosystem Assessment 2003). Examples include plant foods (derived from primary production, pollination, fertilization and seed set), animal foods (derived from secondary production, which in turn relies on primary production etc.), timber for construction (derived from primary production), and a supply of clean water (derived from the collection and filtration of water in watersheds (catchments) and its storage in lakes and streams). Some authors appear to use the terms ecosystem function and ecosystem service interchangeably (e.g., de Groot 1987; Luck et al 2003; Tscharntke et al. 2005), which implies (perhaps correctly) that all ecosystem functions benefit humans in some way.

catchment. The watershed has traditionally been an important scale of analysis in ecosystem ecology (Likens 1992; Groffman et al. 2004) – it is large enough to observe cycles of energy, water and nutrients within the system, but small enough to be relevant for environmental management. A watershed has clear, physical boundaries that define its geographic extent. Within these boundaries, a vast array of biological, chemical and physical processes combine to create cycles of water, carbon and other nutrients, and to capture, transfer and dissipate energy throughout the system (Loreau 2010). The study of ecosystem dynamics is concerned with disturbances or perturbations of ecosystems (both natural and anthropogenic) and their consequences over time and space (Turner et al. 2003; Chapin et al. 2011). Disturbances may include short-term events, such as wildfire, disease, storms and floods, or longer-term processes, such as changes to rainfall and temperature regimes, the clearing of native vegetation and the fragmentation of habitat (White and Pickett 1985; Turner et al 2003). In some cases, a change in the intensity or frequency of natural disturbances (e.g., an increase or decrease in fire interval and/or fire intensity) is sufficient to cause important changes in the structure and function of an ecosystem (Turner et al. 2003; Edwards et al. 2015).

As is the case for population and community ecology (see Chapters 3 and 4), the concepts and theories of ecosystem ecology have largely been developed to describe and extend our understanding of non-urban ecosystems, such as forests, grasslands, lakes, oceans or deserts (Groffman et al. 2004). However, many of these concepts and theories have since been successfully applied to urban systems (Pickett et al. 2001; Grimm and Redman 2004; Groffman et al. 2004; Grimm et al. 2008a,b; Pickett and Grove 2009). Almost all ecosystems are open systems to the extent that they depend on the input of solar energy from an external source – the sun (Odum 1983). However, urban ecosystems tend to be characterized by large-scale inputs of non-solar energy plus water and nutrients imported from outside the system via anthropogenic pathways, then exported to surrounding lands, receiving waters and the atmosphere, also via anthropogenic pathways (Kennedy et al. 2007). At the same time, the use and cycling of endogenous energy, water, carbon and other nutrients are greatly reduced in urban ecosystems. In this chapter, I consider how the biophysical processes of urbanization affect important cycles within ecosystems across a range of spatial scales from the local to the global. I also discuss strategies for mitigating these effects, such as increasing the use of endogenous energy, water and nutrients in cities, and reinstating various ecological processes to promote internal cycling.

5.2 Carbon

5.2.1 Introduction to the carbon cycle

The global carbon cycle is made up of a number of cycles that operate over different spatial and temporal scales. Carbon is exchanged between the atmosphere, living organisms, rocks, sediments, soil, oceans, rivers and lakes through a variety of biogeochemical processes. During photosynthesis, plants and cyanobacteria convert light energy from the sun into chemical energy, extracting carbon dioxide (CO_2) from the atmosphere at a rate of 123±8 PgC/year (95% CI: 102–135 PgC/year, where 1 PgC = 10^{15} grams of carbon; Beer et al. 2010) and releasing oxygen in the process. The chemical energy produced is stored as carbohydrate molecules, synthesized from CO_2 and water, which are then used for growth or to fuel cellular processes. In terrestrial ecosystems, this organic carbon moves up through the food chain from plants to herbivores, carnivores and predators, and also to detritivores and saprophytes that feed upon decomposing plant and animal material. An estimated 1950–3050 Pg of organic carbon is stored in the terrestrial biosphere, which consists of plants, animals and other organisms, both living and dead, above-ground and in soils (Batjes 1996; Prentice et al. 2001). A further ~2000 Pg of organic carbon is stored in wetland and permafrost soils (Bridgham et al. 2006; Tarnocai et al. 2009). Soils also store some inorganic carbon in forms such as calcium carbonate. Organic carbon is released from soils to the atmosphere through the process of respiration, while both organic and inorganic carbon can be dissolved from soils and exported to rivers and the oceans.

Aquatic ecosystems also play an important role in the global carbon cycle, particularly through the rapid exchange of carbon between the atmosphere and the oceans. Atmospheric CO_2 dissolves in the surface layer of the oceans and is then converted to carbonate (dissolved inorganic carbon), before being transformed into organic carbon by marine algae and other plants during photosynthesis. This organic carbon moves through the marine food chain and can be recycled via the decomposition and respiration of marine organisms, or deposited into the deep layer of the ocean where it is stored for extended periods before being circulated back to the surface layer (Ciais et al. 2013). Organic carbon is also released back to the atmosphere via the respiration of marine organisms. The intermediate and deep layers of the oceans store an estimated 37,000 Pg of dissolved inorganic carbon, while the surface layer stores ~900 Pg (Ciais et al. 2013). Dissolved organic carbon in the oceans accounts for a further 700 PgC, while the living marine biota contain 3 Pg of organic carbon (Hansell et al. 2009; Ciais et al. 2013). The Earth's atmosphere currently stores ~832 PgC (Prather et al., 2012; Joos et al. 2013).

In recent centuries, human activities including the burning of fossil fuels (coal, oil and natural gas), deforestation, and the production of cement have increased the concentration of carbon in the Earth's atmosphere (Figure 5.1a). This additional atmospheric carbon is mainly in the form of CO_2 and methane (CH_4), which are known as greenhouse gases because they absorb and emit thermal radiation in the troposphere (the lowest layer of the atmosphere), increasing the temperature of the Earth (Falkowski et al. 2000; IPCC 2014). While approximately 60% of anthropogenic CO_2 emissions has been absorbed by the oceans and terrestrial biosphere, the rest remains in the atmosphere (CSIRO and BoM 2014; Figure 5.1b).

Since the beginning of the industrial revolution in 1750, the global atmospheric concentration of CO_2 has increased by more than 40%, while that of methane has increased by 150% (Ciais et al. 2013; Dlugokencky & Tans 2015). This has significantly changed the Earth's climate, including increased air and oceanic temperatures (global warming), altered rainfall patterns, and an increase in the frequency and severity of extreme weather events such as droughts, floods and hurricanes (CSIRO and BoM 2014; IPCC 2014). Anthropogenic carbon emissions have also had a substantial impact on the oceans. The increasing concentration of carbon in seawater has increased its acidity, with potentially serious consequences for calcifying marine organisms including plankton, corals and molluscs (Doney et al. 2009; Hoegh-Guldberg and Bruno 2010). The pH of surface water in the open

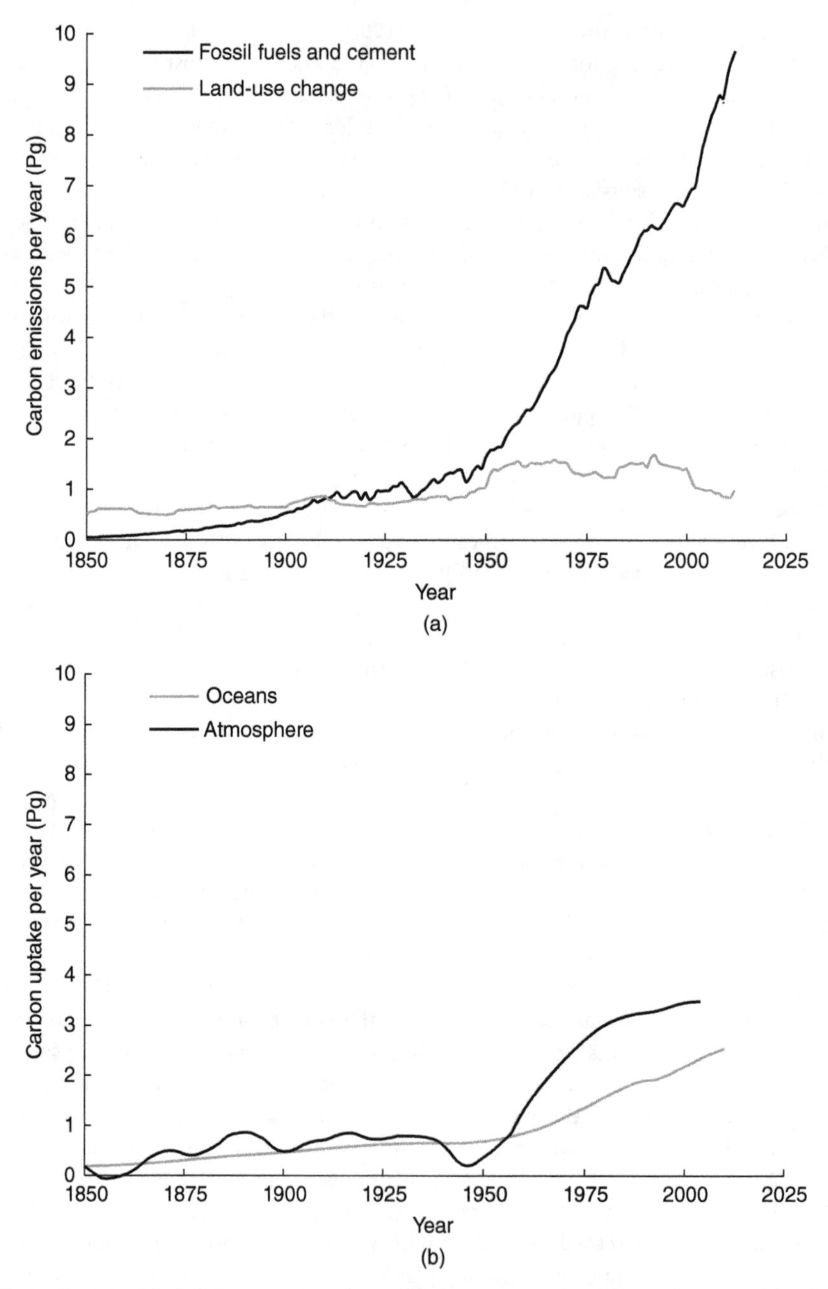

Figure 5.1 (a) Global anthropogenic carbon emissions per year from the combustion of fossil fuels plus cement production (black line) and land-use change (grey line); (b) uptake of carbon emissions per year by the atmosphere (black line) and the oceans (grey line). Data from Le Quéré et al. 2015.

ocean has decreased by 0.1 since 1750, equivalent to a 26% increase in the concentration of hydrogen ions (CSIRO and BoM 2014). Anthropogenic CO_2 emissions continue to rise, and in 2013 were estimated at 36 Pg globally (equivalent to 9.9 ±0.5 PgC; Le Quéré et al. 2015).

5.2.2 Effects of urbanization on the carbon cycle

Urbanization impacts on the carbon cycle in a diversity of ways (Grimm et al. 2008a,b). Cities are energy-intensive ecosystems and their CO_2 emissions correspond quite closely to their energy inputs (Kennedy et al. 2007). Urban areas are estimated to produce more than 80% of the world's anthropogenic CO_2 emissions (Grubler 1994; Svirejeva-Hopkins et al. 2004). Given that approximately 54% of the world's population currently lives in cities (United Nations 2014), this is disproportionately high on a per capita basis. In many parts of the world, power stations burn fossil fuels to provide electricity to cities. This electricity powers homes, industry, office blocks, hospitals, street lights and public transport systems, such as trains and trams. The transport of people and goods into, out of, and around cities by road, sea and air also relies on the combustion of fossil fuels, while oil, gas, coal or wood are commonly burnt to provide winter heating (Kennedy et al. 2009). Greenhouse gas emissions from electricity depend on both the amount of electricity used and the quantity of greenhouse gases emitted for each unit of electricity produced (known as the greenhouse-gas intensity or emissions factor, measured in units of CO_2 equivalent released per unit of energy produced). For example, Cape Town, South Africa draws most of its electricity from coal-fired power stations, and produces 969 g CO_2 eq/kWh (kilowatt hour) of electricity; Geneva, Switzerland, which draws much of its electricity from hydropower, produces 54 g CO_2 eq/kWh (Kennedy et al. 2009).

Increasing urban populations combined with poor planning drive the lateral expansion of cities, increasing the distance that commuters must travel from the outer suburbs to the inner city. An extensive urban form is correlated with higher per-capita energy use in sprawling cities (Kennedy et al. 2007). For example, per-capita CO_2 emissions from Denver, Colorado (population density of 1558 persons/km^2) were 21.5 Mg in 2005, while those from the more compact and densely populated New York City (10,350 persons/km^2) were 10.5 Mg (Hoornweg et al. 2011). A regression analysis of per-capita greenhouse gas emissions from 10 cities reveals that emissions decline as human population density increases (Kennedy et al. 2009; Figure 5.2). However, Denver's higher emissions are partly due to its cooler winters and thus a higher number of heating degree days per year than New York (Kennedy et al. 2009). The expansion of cities also tends to overrun peri-urban agricultural land, which increases the distance that food must be transported from the point of production to the point of consumption by city residents (also known as food miles; Pauchard et al. 2006; van Veenhuizen 2006; Kissinger 2012; Mok et al. 2014).

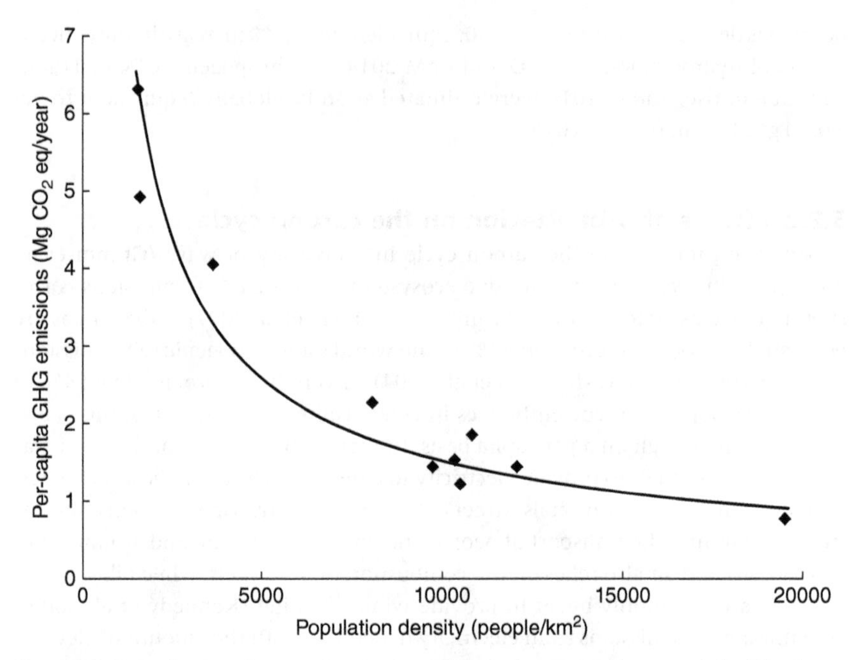

Figure 5.2 Annual per-capita greenhouse-gas emissions from ground transport (in Mg of equivalent CO_2) versus the human-population density of a city. Modified from Kennedy et al. 2009, Figure 3. Reproduced with permission of the American Chemical Society.

The construction and expansion of modern cities drives global demand for cement to construct concrete buildings, roads, pavements, dams, channels, sewers and stormwater-drainage systems. Cement-manufacturing involves heating calcium carbonate to form lime and CO_2. This CO_2 is released to the atmosphere, and the burning of fossil fuels to produce the energy used to manufacture the cement releases further CO_2. As a consequence, the cement industry is responsible for approximately 5% of anthropogenic CO_2 emissions worldwide (range 3–8%), emitting nearly 0.9 Mg of CO_2 for every 1 Mg of cement produced (Mahasenan et al. 2003; Benhelal et al. 2012). China outstrips all other countries in CO_2 emissions from the manufacture of cement clinker and concrete, due to its intensive program of urbanization and the widespread use of concrete in construction (Fernandez 2007). In 2005, total world production of concrete was 2.3 Pg (corresponding to approximately 2 Pg of CO_2 released), with 46% of this attributed to China (Fernandez 2007). Steel is another energy-intensive building material, with an average of 1.8 Mg of CO_2 emitted for every Mg of steel produced (World Steel Association 2013). Steel is used to construct homes, commercial buildings and urban infrastructure, such as railway lines, bridges, power grids, solar panels and wind turbines; it is also used to reinforce concrete. In 2010, the iron and steel industry accounted for approximately 6.7% of total global CO_2 emissions (International Energy Agency, cited by World Steel Association, 2013); global crude steel production reached 1.64 Pg in 2014 (equivalent to 2.95 Pg of CO_2 released; World Steel Association 2014). As for cement, China leads the world in steel production, with 50% of global production in 2014 (World Steel Association 2014).

Urban expansion converts forests, grasslands and agricultural landscapes into urban landscapes, removing vegetation that would otherwise absorb CO_2 from the atmosphere while paving over soils with asphalt and concrete. This conversion produces substantial carbon emissions, estimated at 158 Tg worldwide in 2010 (1 Tg = 10^{12} g; Svirejeva-Hopkins et al. 2004). Urban growth in Africa, China and other parts of the Asia-Pacific made the largest contributions to this figure. The loss of productive agricultural land to urbanization in the USA in the mid-1990s decreased the carbon fixed by photosynthesizing plants by 40 Tg per year (Imhoff et al. 2004). Carbon emissions from urban land conversion are expected to increase further in the coming decades (Svirejeva-Hopkins et al. 2004). However, once urban land conversion has taken place, organic carbon may accumulate in urban and suburban environments that support managed vegetation and soils, such as parks, gardens, street trees, lawns and golf courses (Pataki et al. 2006). While further research is required, results to date suggest that the net effect of urban land conversion on the carbon balance of a city depends partly on the type of ecosystem replaced. For example, the urbanization of an arid, desert environment may increase the carbon stored in vegetation and soils, while the opposite is likely to be true following the urbanization of a humid, forest environment (Golubiewski 2003; Pouyat et al. 2003; Pataki et al. 2006).

Cities and their regions experience elevated concentrations of atmospheric CO_2 and CH_4 (Idso et al. 1998; Kaye et al. 2006; George et al. 2007; Grimm et al. 2008a). The build-up of CO_2 over an urban area resulting primarily from the localized burning of fossil fuels is known as an urban CO_2 dome (Idso et al. 2001). Depending on meteorological conditions, a CO_2 dome may exacerbate the local urban heat-island effect (Balling et al. 2001; George et al. 2007). However, a study in Phoenix, Arizona found that the CO_2 dome over the city was responsible for only a modest proportion of the overall urban heat-island effect (Balling et al. 2001). A combination of global warming from anthropogenic carbon emissions and the urban heat-island effect increased minimum, average and maximum temperatures in Melbourne, Australia by 0.32, 0.23 and 0.14 °C per decade between 1950 and 2005 (Suppiah and Whetton 2007), and decreased average annual frost days from 12.2 (1856–1950) to 3.1 (1951–2000; Figure 5.3; Parris and Hazell 2005). Rising city temperatures increase the need for buildings to be air conditioned during the summer months, which in turn increases electricity consumption and the burning of fossil fuels, creating a positive feedback loop (Baker et al. 2002).

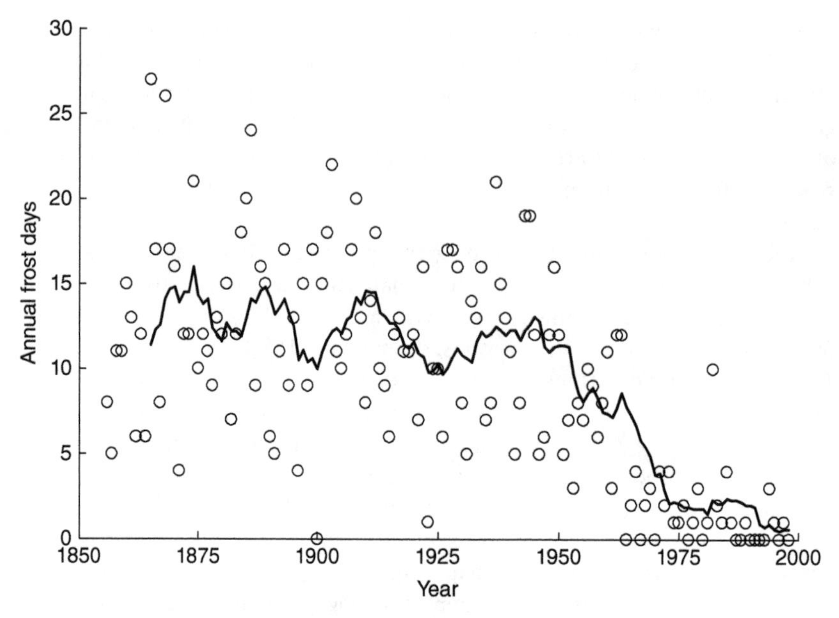

Figure 5.3 Annual frost days in Melbourne, Australia between 1850 and 2000 (circles) with the moving 10-year average of annual frost days (black line). Modified from Parris and Hazell (2005), Figure 3b. Reproduced with permission of Elsevier.

5.2.3 Mitigation strategies

The most obvious way to lessen the impact of urbanization on the global carbon cycle would be to reduce atmospheric carbon emissions associated with the construction, expansion and operation of cities. This could be done by reducing the quantity of fossil fuels burnt to produce electricity and building materials for cities and to power the transport of people and goods; and to a lesser extent by reducing the amount of productive forested and agricultural land that is converted for urban development. However, while conceptually simple, these solutions are difficult to implement in practice. The first international treaty under the United Nations Framework Convention on Climate Change, the Kyoto Protocol (United Nations 1998), was adopted in 1997 and came into force in 2005. Thirty-eight of the developed countries that were party to the protocol committed to reduce their greenhouse gas emissions to an average of 5% below 1990 levels by 2012 (the first implementation period; United Nations 1998), and 21 of these then committed to a further reduction target of at least 18% below 1990 levels by 2020 (the second implementation period; United Nations 2012). In December 2015, 195 countries adopted the first universal climate agreement at the UN Conference on Climate Change in Paris. This agreement aims to reduce net anthropogenic greenhouse gas emissions to zero by the second half of the 21st century, and will become legally binding if signed by at least 55 countries which together represent $\geq$ 55 % of global greenhouse emissions (United Nations 2015).

Cities can reduce their greenhouse gas emissions by reducing demand for electricity and transportation fuel – for example, by increasing the efficiency of industry, reducing per-capita energy usage in households, and reducing the transport of people and goods by road in favour of rail and other public transport systems. Increasing the density of the human population in cities is expected to reduce per-capita energy use, particularly if combined with an efficient public transport system and local planning initiatives that reduce reliance on private vehicles for everyday activities (Kennedy et al. 2009). Cities can also reduce their greenhouse-gas emissions by decreasing the greenhouse-gas intensity of their electricity supply. Coal-powered electricity generation has a median greenhouse-gas intensity of 1001 g CO_2 eq/kWh over its entire lifecycle. This compares to values of 469, 46, 16, 12 and 4 g CO_2 eq/kWh for electricity generation powered by natural gas, photovoltaic solar panels, nuclear reactors, wind turbines and hydroelectricity, respectively (Table 5.1).

Cities that depend on electricity production from coal (e.g., many cities in Australia, China, South Africa and the USA) could substantially reduce their CO_2 emissions by increasing the proportion of electricity produced from other sources, including renewable sources such as solar and wind power. While nuclear power has a low greenhouse-gas intensity, it is not a renewable source of energy. It also poses risks to human health and the environment, as accidents at nuclear power plants can expose workers, the general public and surrounding terrestrial and aquatic habitats to dangerous levels of radioactivity; on average, one such accident occurs every eight years (Maurin 2011). The production of hydroelectricity requires the construction of dams that can have substantial negative impacts on freshwater and terrestrial environments and the livelihood of local people. Dams also pose a risk to human safety if they fail, causing damaging floods. All methods of electricity production impact on humans, other species and the environment (Brook and Bradshaw 2015); a challenge for the future is to choose existing methods and/or develop new methods that have a low greenhouse-gas intensity as well as minimal adverse impacts on other aspects of the environment and human society.

Local actions to reduce carbon emissions include the adoption of green building codes, which can reduce both the embodied energy in buildings (e.g., by using recycled and/or lower-energy materials such as mud bricks, slate and stone) and the amount of energy used for heating, cooling and lighting (e.g., through solar passive design, improved insulation and double-glazing of windows; Reddy and Jagadish 2003; Yudelson 2007; Hammond and Jones 2008; Chua and Oh 2011). Many green building codes or star-rating systems also encourage the local capture of solar or wind power to provide electricity and/or heat water for domestic use (Kajikawa et al. 2011). A lifecycle assessment of three housing styles in Thailand found that a timber house had a greenhouse-gas impact of 18 kg CO_2 eq/m^2 of floor space, compared with 209 kg CO_2 eq/m^2 for a house built with concrete

Table 5.1 The greenhouse-gas intensity of various electricity-generation technologies, in g CO_2 eq/kWh, synthesized from a review of lifecycle analyses of each technology's greenhouse-gas emissions. CCS = carbon capture and storage, PV = photovoltaic, CSP = concentrating solar power; data from Moomaw et al. (2011).

	The greenhouse-gas intensity of various electricity-generation technologies (g CO_2eq/kWh)										
Values	**Bio-power**	**Solar PV**	**Solar CSP**	**Geothermal**	**Hydropower**	**Ocean energy**	**Wind energy**	**Nuclear energy**	**Natural gas**	**Oil**	**Coal**
Minimum	−633	5	7	6	0	2	2	1	290	510	675
25th percentile	360	29	14	20	3	6	8	8	422	722	877
50th percentile	18	46	22	45	4	8	12	16	469	840	1001
75th percentile	37	80	32	57	7	9	20	45	548	907	1130
Maximum	75	217	89	79	43	23	81	220	930	1170	1689
CCS min	−1368								65		98
CCS max	−594								245		396

blocks. A proposed house with fibre-cement panels for walls was calculated to have an intermediate greenhouse gas impact of 140 CO_2 eq/m^2 (O'Brien and Hes 2008). Green building initiatives in Malaysia have led to the construction of a number of energy-efficient office buildings, such as the Ministry of Energy, Green Technology and Water's Low Energy Office Building in Putrajaya and the Malaysia Green Technology Corporation's Green Energy Office Building in Bandar Baru Bangi (Chua and Oh 2011). These two buildings have a building energy index (BEI) of 100 and 35 kWh/m^2/year, respectively, compared to an average BEI of 250 kWh/m^2/year for conventional office buildings in Malaysia (Chua and Oh 2011).

Urban vegetation and soils play an important role in sequestering carbon from the atmosphere and storing it in above- and below-ground biomass (Jo and McPherson 1995; Nowak and Crane 2002; Pataki et al. 2006; Stoffberg et al. 2010). Trees, gardens, green roofs and green walls also mitigate the local effects of global warming and the urban heat island in temperate and tropical cities through shading and increased evapotranspiration (Tyrvainen et al. 2005; Roth 2007; Norton et al. 2013). Considering seven cities in the USA, a single street or park tree reduces CO_2 emissions by 31–252 kg/year through a combination of sequestration and avoidance (the reduction in emissions gained from energy savings associated with tree cover; Pataki et al. 2006). Scaling up to an entire city, the street and park trees of Glendale, Arizona and Minneapolis, Minnesota were estimated to reduce CO_2 emissions by 0.7 Gg and 50 Gg per year in each city, respectively (1 Gg = 1 billion g; McPherson et al. 2005; Pataki et al. 2006). By area, reductions in CO_2 emissions are estimated at 45.8 kg/ha/year in Glendale and 4118 kg/ha/year in Minneapolis. Maintaining or increasing green open space with permeable soils, planting broad-leaved trees in streets, parks and private gardens to provide summer shade, creating green roofs and walls on existing buildings, and irrigating these various types of vegetation with rainwater or stormwater runoff to keep them growing actively and thus absorbing CO_2 from the atmosphere during dry weather (see Section 5.3.3) would all contribute to the dual goals of carbon sequestration and local cooling (Oberndorfer et al. 2007; Norton et al. 2013, 2015). For example, the shading of buildings by street trees in Berkeley, California saves an estimated 95 kWh of electricity/tree (~3.5 million kWh across the whole city) during summer from avoided use of electricity for cooling (McPherson et al. 2005).

Further benefits of urban vegetation include the removal of pollutants from the atmosphere (e.g., ozone, CO, SO_2, NO_2 and particulates), uptake of pollutants from contaminated soils, and provision of habitat for local biodiversity (Dwyer et al. 1992; Dimoudi and Nikolopoulou 2003; McPherson et al. 2005; Nowak et al. 2006; Oberndorfer et al. 2007). One potential disadvantage of urban trees is the production of biogenic volatile organic compounds (BVOCs) by certain species, which may contribute to the secondary formation of ground-level ozone if other

Figure 5.4 A garden allotment in York, UK, growing vegetables and berries. Photograph by Kirsten M. Parris.

precursor pollutants are present in sufficiently high concentrations (McPherson et al. 2005; Calfapietra et al. 2013). This problem can be solved by planting species that are low or non-emitters of BVOCs (Calfapietra et al. 2013). Depending on the species planted, urban vegetation can also contribute to local food production in cities. Urban agriculture (or urban farming) reduces the distance that food must travel from its site of production to city residents (food miles), thus reducing the use of fossil fuels for cold storage and transport (Smit and Nasr 1992; McClintock 2010). While urban agriculture is increasing in popularity in developed countries, such as Australia, the UK and the USA (Lovell 2010; McClintock 2010; Mok et al. 2014; Figure 5.4), it already represents an important source of fresh food for city dwellers in many parts of Africa, Asia and Eastern Europe (Mougeot, 2005). Cuba leads the world in urban farming, with a 1000-fold increase in urban agricultural production since 1994 (Koont 2011). Following the collapse of the Soviet Union, Cuba could not access the fossil fuels and fertilisers needed for large-scale, industrial agriculture. It therefore refocused its agricultural programs towards sustainable organic farming on disused public land in and around cities, with substantial success.

5.3 Water

5.3.1 Introduction to the water cycle

Water moves in a continuous cycle between reservoirs in the atmosphere, the biosphere, the ground, the oceans, ice sheets and glaciers, and bodies of liquid freshwater such as streams, rivers and lakes. Like the carbon cycle, the water cycle includes many processes that operate over a range of temporal and spatial scales. Heat from the sun causes water to evaporate and enter the atmosphere as vapour. This atmospheric water vapour is moved by wind currents, rises, cools and falls as precipitation (rain, sleet or snow). Most of the rain that falls on land infiltrates the soil, where it may be stored as soil moisture, percolate through to streams, ponds or lakes via subsurface flow, or enter underground aquifers; a small proportion flows overland to streams and other water bodies. Average global precipitation is estimated to be around 0.5 Pm^3/year (1 $Pm^3 = 10^{15}$ cubic metres; Pidwirny 2006). Approximately 80% of this falls over the oceans and the remaining 20% on land. In terrestrial environments, plants take up moisture from the soil and transfer it back to the atmosphere via the process of evapotranspiration. Globally, the volume of evapotranspiration over the vegetated land surface of Earth is estimated to be 62.8 Tm^3/year (Mu et al. 2011), >60% of terrestrial precipitation. Streams and rivers flow to the sea, returning rainwater to the oceans, while lakes and underground aquifers may store water for longer periods. Water can also be stored for extended periods – up to hundreds of thousands of years – as ice and snow before melting or subliming (changing directly from solid water to vapour).

5.3.2 Effects of urbanization on the water cycle

Urbanization affects the water cycle through land-cover change (removal of existing vegetation; construction of buildings, roads and other urban infrastructure; replacement of permeable surfaces and flow paths with hard surfaces and lined drains; reduction in the area of open space; and modification or destruction of aquatic habitats) and the production of pollution and waste. Impermeable surfaces prevent the infiltration of rainwater into soils (Stone 2004). When combined with efficient stormwater drainage systems, this water is instead transported rapidly to streams, lakes and rivers (Leopold 1968; Walsh et al. 2005). Destruction of ephemeral ponds or soaks and the modification of natural drainage lines (e.g., the replacement of ephemeral, first-order streams with concrete or metal piping, and/or the constriction of higher-order streams into straight, concrete-lined channels) also increase the speed with which water flows following a rainfall event, further compounding the problem (Lindh 1972; Walsh et al. 2005). In addition, removal of vegetation and the loss of open, permeable space reduce the rate of evapotranspiration in cities – rather than trees and other plants taking up water from the soil and releasing it into the atmosphere, there is less permeable soil surface available to absorb water and there are many fewer plants available to transpire it (Kondoh and Nishiyama 2000; Walsh et al. 2012).

These changes in urban environments increase the local frequency and magnitude of erosive flows of water, increase peak flows in urban streams, reduce lag times between a rainfall event and the arrival of peak flows, and cause a rapid rise and fall of the storm hydrograph (the "flashy flow" of urban streams), even after relatively small rainfall events (Paul and Meyer 2001; Stone 2004; Walsh et al. 2005; Chapter 2). Considering the ecosystem of a city as a whole, the import of water for drinking, domestic and industrial uses, the export of much of this water through the sewage (wastewater) system, and the rapid egress of rainwater from the urban environment via the stormwater system tend to dominate urban water flows (Figure 5.5). This undermines the cyclical nature of the urban water cycle; instead, the modern city acts as the arrival and departure point for large volumes of water travelling on two mostly separate paths – potable water flowing into the city and then out again via the sewage system, and rainwater falling on the city and then flowing out via the stormwater system.

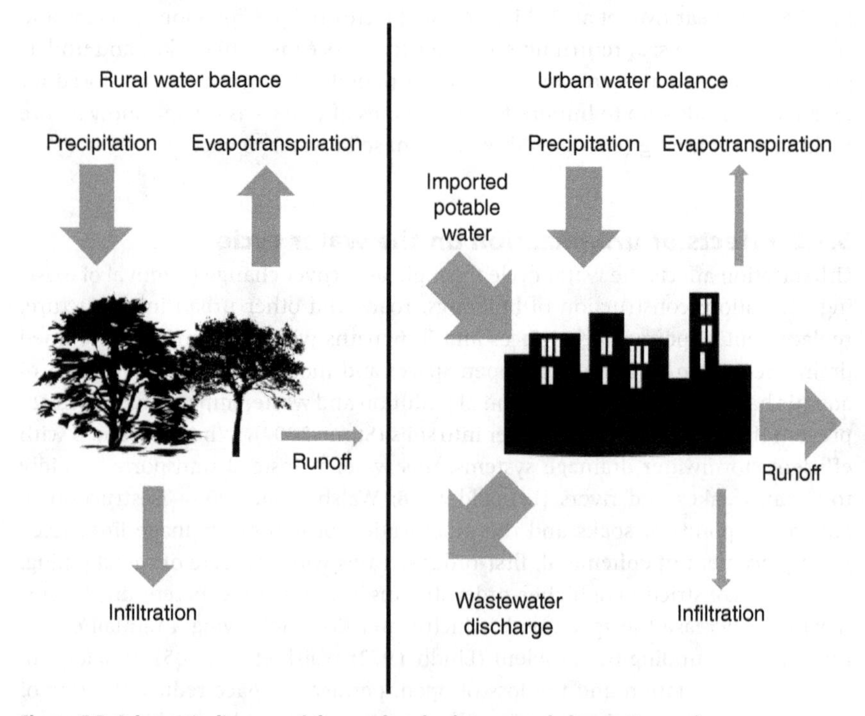

Figure 5.5 Schematic diagram of the rural and urban water balance. Rates of evapotranspiration and infiltration are lower per unit area in urban than rural ecosystems, while the volume of runoff is higher. Urban ecosystems also tend to be characterized by the import of large volumes of potable water and the export of large volumes of wastewater.

In formal settlements within most modern cities, sewage wastewater is managed in central treatment plants, reducing the load of nutrients, chemicals, microbial pathogens and other pollutants released to streams and coastal waters. However, urban stormwater runoff arrives in streams and other receiving waters untreated, carrying with it a host of pollutants including nitrogen, phosphorus, petroleum products, heavy metals and faecal contamination from pet droppings (Lee et al. 2002; Hatt et al. 2004; Walsh et al. 2005, 2012; Li et al. 2007; Imberger et al. 2011). It should be noted that in some older cities in Europe and North America, the sewage and stormwater systems are less clearly separated, magnifying pollution events in receiving waters following rainfall (Grimm et al. 2008a). The problem of urban water pollution is even more significant in cities with large informal settlements or slums. Informal settlements are often built without planning controls, sometimes in areas subject to flooding, and lack basic sanitation and infrastructure to manage stormwater. Following rainfall, these conditions lead to the co-mingling of stormwater runoff, human excreta and the droppings of domestic animals such as poultry, goats, dogs and cattle, spreading a range of pathogens, causing disease and polluting receiving waters (Jagals 1996; Parkinson 2003; Owusu-Asante and Stephenson 2006; Owusu-Asante and Ndiritu 2009). These receiving waters are then often used for other human purposes, including drinking, fishing and the irrigation of crops (Jagals 1996; Parkinson et al. 2007). Standing, contaminated water in open drains or soaks also poses a serious health risk to the inhabitants of informal settlements, and provides a breeding ground for disease-transmitting mosquitoes and parasitic worms (Parkinson 2003).

5.3.3 Mitigation strategies

The challenges of urban water management arise from the difficulty of managing potable water, wastewater and stormwater to achieve multiple – and sometimes competing – objectives (Grimm et al. 2008a). Efficient management of wastewater in enclosed pipes or sewers that flow to designated treatment plants, combined with efficient drainage of stormwater runoff from impermeable urban surfaces, meet the objectives of sanitation, disease-prevention, flood-prevention, and the efficient movement of water-borne pollutants from urban streets into streams and other receiving waters. However, these practices also degrade urban streams, transfer pollutants to downstream waterways, and reduce opportunities to retain and use stormwater in ways that can improve the urban environment (e.g., by supporting vegetated systems such as parks, street trees and green roofs, which in turn provide a range of ecosystem services such as shading, microclimatic cooling, absorption of airborne pollutants and provision of habitat for native biodiversity; Paul and Meyer 2001; Wong 2007; Fletcher et al. 2007; Ashley et al. 2013; Fletcher et al. 2014).

Urban water management has become more complicated in recent decades, with an increasing number of stated objectives (Figure 5.6). In developed countries, growing realization of the ecological and social benefits of retaining

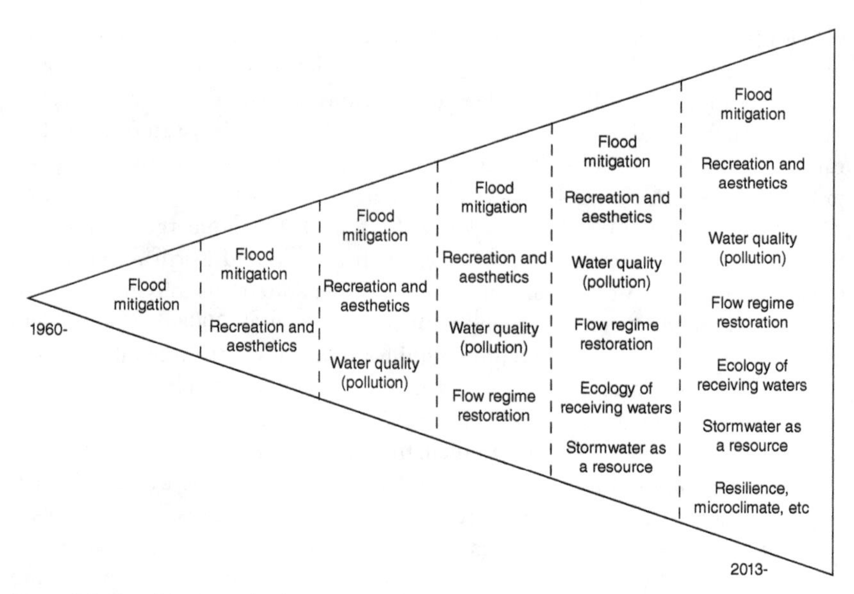

Figure 5.6 The objectives of urban water management versus time. Fletcher et al. (2014), Figure 2.

stormwater in urban catchments has spawned a new approach to urban water management that is variously known as Water-Sensitive Urban Design (Australia), Low-Impact Design, Green Infrastructure (USA), Sustainable Urban Drainage Systems (UK) and Techniques Alternatives (France), among others (Fletcher et al. 2014). Water-Sensitive Urban Design and related practices aim to protect and enhance natural water systems in urban areas; integrate stormwater treatment into the landscape; reduce stormwater runoff and peak stream flows via local detention measures and a decrease in connectedness between impervious surfaces and receiving waters; and improve the quality of water draining from urban areas (Fletcher et al. 2014, and references therein). These practices employ a range of engineering and planning measures to reduce the volume of stormwater that reaches urban streams, the rate at which it arrives following rainfall events, and the load of pollutants it carries (Wong 2007). Examples include the installation of water tanks, rain gardens, permeable pavements, green roofs, retention ponds, constructed wetlands and earthen swales (Walsh et al. 2005; Mentens et al. 2006; Hilten et al. 2008; Beecham et al. 2012).

Although the principles of Water-Sensitive Urban Design are likely to apply in all types of cities, current water-management priorities in many informal settlements remain the prevention of flooding and the effective separation of wastewater from stormwater (Parkinson et al. 2007; Okoko 2008; Owusu-Asante and Ndiritu 2009). Until these issues are addressed, management objectives such as the restoration of natural flow regimes, the ecology of receiving waters, and recreation and aesthetics (Figure 5.6) are unlikely to be given much weight. Some of the techniques for urban water management outlined above are appropriate for slums and other informal settlements, particularly the use of permeable pavements and the harvesting of rainwater to be used as a potable water supply for residents (Handia et al. 2004; Cowden et al. 2006; Owusu-Asante and Ndiritu 2009). However, the desirability of retaining stormwater in urban catchments via structures such as retention ponds, constructed wetlands and swales is negated when this stormwater is contaminated with human waste – any measures to ensure such water drained more slowly from the urban environment would pose a risk to human health (Parkinson 2003). If human waste can be effectively managed in a separate, closed system, a range of engineering measures to reduce the flow and nutrient load of stormwater from informal settlements is likely to be both practical and advantageous.

5.4 The nitrogen cycle

Seventeen nutrients are known to be essential for plant function and growth; carbon, hydrogen, oxygen, nitrogen, phosphorus, potassium, calcium, magnesium, sulfur, boron, chlorine, copper, iron, manganese, molybdenum, nickel and zinc (Barker and Pilbeam 2015). Plants take up carbon from the atmosphere in the form of CO_2, and hydrogen and oxygen from water. They obtain the other 14 essential nutrients from the mineral component of the soil. Plant growth – and thus the primary productivity of ecosystems – is often limited by the availability of nutrients, including the concentration of carbon in the atmosphere and of nitrogen, phosphorus and other mineral nutrients in the soil. Thus, enhanced atmospheric CO_2 may increase the rate of photosynthesis and plant growth, assuming other nutrients are not limiting (Wang et al. 2012; Reich and Hobbie 2013). The availability of nitrogen and phosphorus also influences the rate at which organic matter decomposes in soils, humus, leaf litter and water (Flanagan and Van Cleve 1983; Aerts 1997; Manzoni et al. 2010). Every essential nutrient has its own biogeochemical cycle, and there is insufficient space in this chapter to discuss them in detail. However, each of these cycles is likely to be affected by the construction and expansion of cities. As the global human population becomes increasingly urbanized, food and energy resources will become increasingly concentrated in cities. Urban areas are therefore expected to play an important role in the global distribution of nutrients (Bernhardt et al. 2008).

Nitrogen is one of the most important nutrients for all organisms, often limiting plant growth and primary productivity (Vitousek et al. 1997; Dentener et al. 2006; Gruber and Galloway 2008; Zehr and Kudela 2011). As an essential component of proteins and the nucleic acids DNA and RNA, nitrogen is found in every living cell. Plants and blue-green algae also require nitrogen to produce chlorophyll, the green pigment that plays a key role in photosynthesis. Nitrogen (as N_2) is the dominant gas in the Earth's atmosphere and comprises the largest pool of nitrogen in the global cycle. However, this unreactive nitrogen is not available for uptake by plants, and without human intervention must first be converted to ammonia (NH_4^+) by diazotrophic bacteria or to nitrogen oxides (NO_x) by lightning. These processes are known as nitrogen fixation. Diazotrophs occur either as free-living organisms (e.g., certain anaerobic bacteria in soils and ocean sediments) or in symbiosis with plants (e.g., rhizobia in the root nodules of legumes). They fix nitrogen by combining gaseous N_2 with hydrogen to produce ammonia. The oxidation of N_2 by lightning during electrical storms produces reactive nitrogen oxides. These combine with hydroxyl radicals (OH) to form nitric acid (HNO_3), which is then transferred to terrestrial, freshwater or marine environments on the Earth's surface via precipitation (Bond et al. 2002).

In terrestrial habitats, plants assimilate nitrogen via their roots as ammonium or nitrate ions, before incorporating them into organic compounds required for function and growth (e.g., chlorophyll, amino acids and nucleic acids). Animals, in turn, satisfy most of their nitrogen requirements by consuming plants and/or other animals. When an organism dies or an animal excretes waste, the incorporated organic nitrogen is converted to ammonia by bacteria or fungi in a process known as ammonification. Other groups of bacteria convert ammonia to nitrite and then nitrate (nitrification). A further group of bacteria converts nitrates back to N_2 (denitrification) under anaerobic conditions such as those found in waterlogged soil, thereby completing the nitrogen cycle (Gruber and Galloway 2008). Similar processes occur in the marine component of the nitrogen cycle, with blue-green algae (cyanobacteria) responsible for the majority of nitrogen fixation. In coastal waters and the open ocean, phytoplankton assimilate nitrogen as ammonium or nitrate, and the concentration of bio-available nitrogen is a key driver of primary and secondary production (Gruber and Galloway 2008; Zehr and Kudela 2011).

5.4.1 Effects of urbanization on the nitrogen cycle

Human activities – particularly the combustion of fossil fuels, industrial fixation of N_2 via the Haber–Bosch process (to create ammonia-based fertilisers), and the cultivation of crops such as legumes and rice that are associated with nitrogen fixation by free-living and symbiotic bacteria – have greatly enhanced the availability of nitrogen (Galloway et al. 2004). The global rate of anthropogenic nitrogen fixation (210 Tg/yr of reactive nitrogen) is 10 times higher than a century ago, and is now equal to nitrogen fixation from natural sources (Fowler et al. 2013). Given the high concentration of the world's population, energy and nutrients in cities, much of this reactive nitrogen is either produced in urban environments (through the combustion of fossil fuels for electricity generation, heating and vehicular transport) or brought to cities in the form of food, garden fertiliser and industrial chemicals (Bernhardt et al. 2008; Figure 5.7).

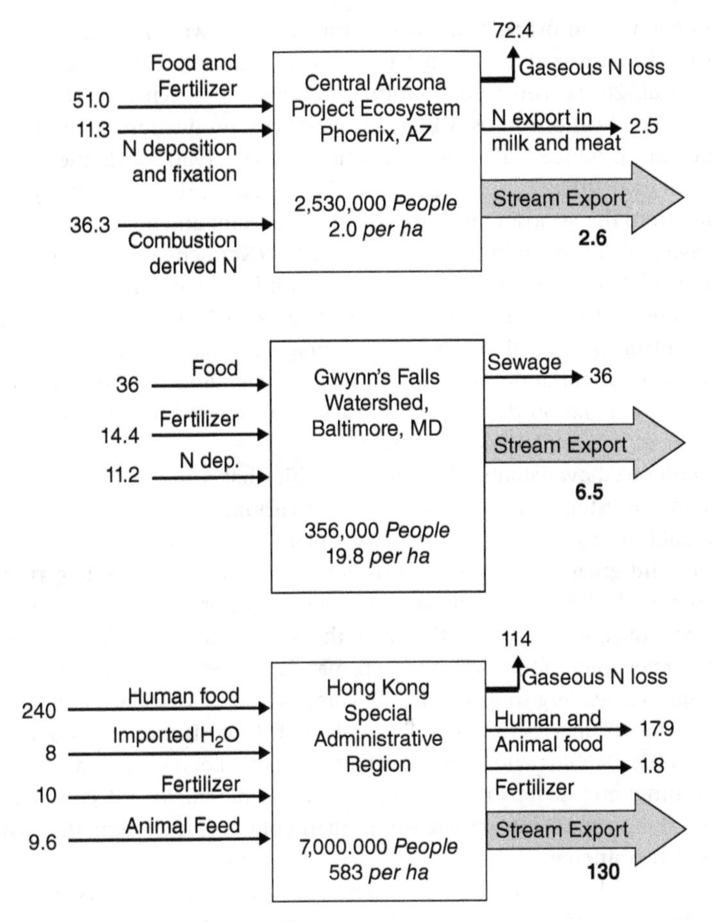

Figure 5.7 The nitrogen mass balance of three cities, showing inputs and outputs of nitrogen: Phoenix, Arizona, USA; Gwynn's Falls Watershed, Baltimore, USA; and the Hong Kong Special Administration Area. Bernhardt et al. 2008, figure 2. Reproduced with permission of John Wiley & Sons.

In cities with sufficient infrastructure and well-developed environmental regulations, most of the nitrogen imported as human food and industrial compounds is contained and treated within septic or sewage systems or disposed of as solid waste in designated ground-fill sites (Bernhardt et al. 2008). However, nitrogen that arrives in cities in the form of pet food and garden fertilizer is not so well controlled. As a consequence, pet waste and the runoff from fertilized lawns and garden beds make a significant contribution to the N balance of urban ecosystems (e.g., Baker et al. 2001; Groffman et al. 2004). For example, an estimated 58 and

14% of N inputs in urban Maricopa County, Arizona were from garden fertilizer and pet waste, respectively, with a total contribution of 10.4 and 2.4 Gg N/year (Baker et al. 2001). Nitrogen is emitted in vehicular exhaust as NO_x and NH_3, while the combustion of fossil fuels for electricity production, domestic heating and industry produces substantial quantities of NO_x (van Aardenne et al. 2001; Bernhardt et al. 2008; Gu et al. 2012, 2013). Localized, wet and dry deposition of nitrogen from the atmosphere comprises an important and increasing component of the available nitrogen in urban environments (Kennedy et al. 2007). In China, emissions of NO_x from the combustion of fossil fuels have increased more than six-fold since 1980, to an estimated 6 Tg N/year in 2008 (Gu et al. 2012). The average nitrogen deposition rate across China is estimated at 12.9 kg N/ha/year, with substantial variation between provinces depending on the level of urbanization and population density (range: 3.8 kg N/ha/year in Xinjiang to 38.2 kg N/ha/year in Taiwan; Lu and Tian 2007).

The enhanced availability of nitrogen in cities has a number of consequences for urban ecosystems. In terrestrial urban environments, an increased supply of bio-available nitrogen may lead to faster plant growth, increased primary productivity and greater accumulation of biomass than in surrounding rural areas (Vitousek et al., 1997). In regions where the local flora are adapted to soils low in nitrogen (e.g., Australia, South Africa, the Mediterranean, California and other desert biomes), additional soil nitrogen may favour exotic, invasive plants over native species, leading to a shift in the composition of vascular plant communities and a loss of species richness (Bobbink et al. 1998; Dukes and Mooney 1999; Brooks 2003; Daehler 2003). Similar patterns have been observed for ectomycorrhizal fungi and lichens near Oulu, northern Finland, with decreasing species richness along a gradient of increasing urban pollution (including increasing soil nitrogen; Tarvainen et al. 2003).

The increased availability of nitrogen in urban environments also impacts on freshwater and marine systems, sometimes quite dramatically. Efficient stormwater-drainage networks that rapidly deliver nitrogen from the terrestrial urban environment to receiving waters such as streams, lakes, estuaries and oceans following rainfall events, outfalls of sewage (both treated and untreated) and the deposition of nitrogen from the polluted urban atmosphere all contribute to elevated nitrogen levels in aquatic habitats (Rabalais 2002; Bernhardt et al. 2008). Consequences include increased primary production, eutrophication (an accumulation of organic matter), increased turbidity and a corresponding reduction in the amount of sunlight reaching submerged aquatic plants, harmful algal blooms, oxygen depletion, a loss of plant diversity and the collapse of fish populations (Rabalais 2002; Camargo et al. 2005). (However, it is worth noting that the productivity of some streams is limited more by the availability of phosphorus than of nitrogen, and an influx of P from urbanized watersheds may have a greater impact on in-stream processes such as eutrophication than an influx of N (e.g., Taylor et al. 2004)). Nitrate is toxic to a variety of freshwater animals, including invertebrates, fish and amphibians; toxicity increases with nitrate concentration and exposure times, but decreases with body size and salinity (Camargo et al. 2005). A recommended safe level of nitrate is 2 mg NO_3-N/l in freshwaters and 20 mg NO_3-N/l in marine systems (Camargo et al. 2005). Deposition of ammonium and nitrate can also increase the acidity of freshwater systems, which in turn reduces the survival, size and abundance of fish (Rabalais 2002). Lastly, nitrogen deposition in rivers, estuaries and coastal waters increases emissions of nitrous oxide (N_2O) from these water bodies to the atmosphere (Seitzinger and Kroeze 1998). Nitrous oxide is a potent greenhouse gas with a global warming potential >200 times greater than that of CO_2 over a time horizon of 20 years (Ciais et al. 2013).

5.4.2 Mitigation strategies

There are a number of possible strategies to reduce the availability of reactive nitrogen in urban and adjacent environments, some more difficult to implement than others. As for carbon emissions, these include decreasing emissions of NO_x and NH_3 in urban and adjacent environments by reducing the quantity of fossil fuels (particularly coal and oil) burnt to provide electricity, transport, heating and building materials for cities (see Section 5.2.3). Tighter environmental regulation of NO_x emissions from coal-fired power stations, particularly in developing countries, could also make an important difference to global NO_x pollution. For example, NO_x pollution from the combustion of fossil fuels in China is currently loosely regulated; these emissions continue to rise, and are substantially higher per unit of energy produced than in Europe and North America (Gu et al. 2013; Hill 2013). In 2010, emissions of NO_x from China relative to GDP were 3.4 Mg/million dollars, compared with 0.9 Mg NO_x/million dollars of GDP in the USA and 0.3 Mg NO_x/million dollars of GDP in Japan (Hill 2013). However, other countries such as Turkey and Australia also had relatively high NO_x emissions based on this measure (2.3 and 1.8 Mg NO_x/million dollars of GDP, respectively; Hill 2013). Nitrogen deposition may be further reduced in urban areas by imposing stricter standards on the emissions of NO_x and NH_3 from motor vehicles and regulating atmospheric concentrations of NH_3 (Moomaw 2002; Bernhardt et al. 2008).

Extending sanitary infrastructure in cities and applying advanced wastewater treatment methods would substantially reduce the export of nitrogen from urban areas to receiving waters (Bernhardt et al. 2008). For example, one third of the population of the European Union had no sewage treatment facilities at all in 2005, although substantial progress had been made over the preceding decade in countries such as Romania, Cyprus, Slovenia, Iceland and Spain (Katsiri 2009). Measures to reduce the speed and volume of stormwater drainage and to retain this water in the urban landscape will also reduce nitrogen export to streams and other receiving waters (see Section 5.3.3). These include the diversion of stormwater to household tanks and the retention and denitrification of stormwater in vegetated wetlands, swales and retaining ponds (Hsieh et al. 2007; Bernhardt et al. 2008). Imports of nitrogen to urban environments could be reduced through guidelines or regulations to limit the use of inorganic nitrogen fertilisers on sports fields, lawns, parks and gardens, while stricter guidelines and/or appropriate enforcement of existing guidelines regarding the responsible disposal of pet wastes would also reduce the export of reactive nitrogen from cities and towns (Driscoll et al. 2003).

5.5 Summary

The construction, expansion and operation of cities have substantial impacts on ecosystems across a range of spatial scales, from local ecosystem processes and functions to global cycles of water, carbon and other nutrients. The continued movement of the world's human population from rural to urban areas will see an increasing concentration of global energy, water, nutrients and other resources in cities over time. However, there are many opportunities to reduce the adverse impacts of urbanization on ecosystems by changing practices such as the large-scale combustion of fossil fuels for electricity production, domestic heating and the manufacture of building materials; the construction of energy-inefficient houses and office buildings; the reliance on private vehicles for the transport of people and goods; the loss of productive farmland in peri-urban areas; the removal of transpiring vegetation from cities; engineering practices that lead to the rapid export of stormwater and the pollutants it carries to local streams and other receiving waters; and widespread use of inorganic nitrogenous fertilisers in urban parks and gardens. Through increasing the proportion of electricity generated from renewable sources, such as wind and solar power; tighter regulation of pollutants arising from the combustion of fossil fuels; improved public-transport systems; adoption of green building practices, urban greening strategies, water-sensitive urban design, and measures to limit continuing urban sprawl; the cities of the future will be able to support more people with fewer negative impacts on ecosystem processes, functions and services. In the next chapter, I address the ecology of human populations in cities, and the many ways in which urbanization affects human health and wellbeing.

Study questions

1 What is an ecosystem?
2 Define the terms ecosystem process, ecosystem function and ecosystem service. How does each relate to the other?
3 Explain the principal ways in which urbanization affects the global carbon cycle, and practical strategies for mitigating these effects.
4 Do you think that developed countries have an extra responsibility to reduce the impacts of their cities on the global carbon and water cycles?

5 Outline the principles of water-sensitive urban design. How can these principles be applied in formal and informal settlements?
6 The outward expansion of cities should be limited to protect productive farm land and decrease per-capita energy use – discuss.
7 Describe the ecological impacts of excess nitrogen in urban environments (terrestrial and aquatic). How would you reduce these in a town or city that you know?

References

Aerts, R. (1997) Climate, leaf litter chemistry and leaf litter decomposition in terrestrial ecosystems: a triangular relationship. *Oikos*, **79**, 439–449.

Ashley, R., Lundy, L., Ward, S. *et al.* (2013) Water-sensitive urban design: opportunities for the UK. *Proceedings of the ICE-Municipal Engineer*, **166**, 65–76.

Baker, L.A., Brazel, A.J., Selover, N. *et al.* (2002) Urbanization and warming of Phoenix (Arizona, USA): impacts, feedbacks and mitigation. *Urban Ecosystems*, **6**, 183–203.

Baker, L.A., Hope, D., Xu, Y. *et al.* (2001) Nitrogen balance for the Central Arizona-Phoenix (CAP) ecosystem. *Ecosystems*, **4**, 582–602.

Balling, R.C., Cerveny, R.S., and Idso, C.D. (2001) Does the urban CO_2 dome of Phoenix, Arizona contribute to its heat island? *Geophysical Research Letters*, **28**, 4599–4601.

Barker, A.V. and D.J. Pilbeam (eds). 2015. *Handbook of Plant Nutrition*, 2nd edn, CRC Press, Boca Raton.

Batjes, N.H. (1996) Total carbon and nitrogen in the soils of the world. *European Journal of Soil Science*, **47**, 151–163.

Beecham, S., Kandasamy, J., and Pezzaniti, D. (2012) Stormwater treatment using permeable pavements. *Water Management*, **165**, 161–170.

Beer, C., Reichstein, M., Tomelleri, E. *et al.* (2010) Terrestrial gross carbon dioxide uptake: global distribution and covariation with climate. *Science*, **329**, 834.

Benhelal, E., Zahendi, G., and Hashim, H. (2012) A novel design for green and economical cement manufacturing. *Journal of Cleaner Production*, **22**, 60–66.

Bernhardt, E.S., Band, L.E., Walsh, C.J. *et al.* (2008) Understanding, managing, and minimizing urban impacts on surface water nitrogen loading. *Annals of the New York Academy of Sciences*, **1134**, 61–96.

Bobbink, R., Hornung, M., and Roelofs, J.M. (1998) The effects of airborne pollutants on species diversity in natural and semi-natural European vegetation. *Journal of Ecology*, **86**, 717–738.

Bond, D.W., Steiger, S., Zhang, R. *et al.* (2002) The importance of NOx production by lightning in the tropics. *Atmospheric Environment*, **36**, 1509–1519.

Bridgham, S.D., Megonigal, J.P., Keller, J.K. *et al.* (2006) The carbon balance of North American wetlands. *Wetlands*, **26**, 889–916.

Brook, B.W. and Bradshaw, C.J.A. (2015) Key role for nuclear energy in global biodiversity conservation. *Conservation Biology*, **29**, 702–712.

Brooks, M.L. (2003) Effects of increased soil nitrogen on the dominance of alien annual plants in the Mojave Desert. *Journal of Applied Ecology*, **40**, 344–353.

Calfapietra, C., Fares, S., Manes, F. *et al.* (2013) Role of Biogenic Volatile Organic Compounds (BVOC) emitted by urban trees on ozone concentration in cities: A review. *Environmental Pollution*, **183**, 71–80.

Camargo, J.A., Alonso, A., and Salamanca, A. (2005) Nitrate toxicity to aquatic animals: a review with new data for freshwater invertebrates. *Chemosphere*, **58**, 1255–1267.

Chapin, F.S., Matson, P.A., and Mooney, H.A. (2011) *Principles of Terrestrial Ecosystem Ecology*, Springer, New York.

Chua, S.C. and Oh, T.H. (2011) Green progress and prospect in Malaysia. *Renewable and Sustainable Energy Reviews*, **15**, 2850–2861.

Ciais, P., Sabine, C., Bala, G. *et al.* (2013) Carbon and other biogeochemical cycles, in *Climate Change 2013: The Physical Science Basis. Contribution of Working Group I to the Fifth Assessment Report of the Intergovernmental Panel on Climate Change* (eds T.F. Stocker, D. Qin, G.K. Plattner, *et al.*), Cambridge University Press, Cambridge, pp. 465–570.

Costanza, R.R., d'Arge, R., de Groot, R.S. *et al.* (1997) The value of the world's ecosystem services and natural capital. *Nature*, **387**, 253–260.

Commonwealth Scientific and Industrial Research Organisation (CSIRO) and Australian Bureau of Meteorology (BoM). 2014. *State of the Climate 2014*. Australian Government Bureau of Meteorology, Canberra. http://www.bom.gov.au/state-of-the-climate (accessed 23/06/2014)

Cowden, J.R., Mihelcic, J.R., and Watkins, D.W. (2006) Domestic rainwater harvesting assessment to improve water supply and health in Africa's urban slums, in *Proceedings of the World Environmental and Water Resource Congress 2006: Examining the Confluence of Environmental and Water Concerns* (ed R. Graham), American Society of Civil Engineers, Reston, pp. 1–10.

Crutsinger, G.M., Collins, M.D., Fordyce, J.A. *et al.* (2006) Plant genotypic diversity predicts community structure and governs an ecosystem process. *Science*, **313**, 966–968.

Daehler, C.C. (2003) Performance comparisons of co-occurring native and alien invasive plants: implications for conservation and restoration. *Annual Review of Ecology, Evolution, and Systematics*, **34**, 183–221.

de Groot, R.S. (1987) Environmental functions as a unifying concept for ecology and economics. *Environmentalist*, **7**, 105–109.

Dentener, F., Drevet, J., Lamarque, J.F. *et al.* (2006) Nitrogen and sulfur deposition on regional and global scales: A multimodel evaluation. *Global Biogeochemical Cycles* **20**, GB4003, doi:10.1029/2005GB002672.

Diaz, S. and Cabido, M. (1997) Plant functional types and ecosystem function in relation to global change. *Journal of Vegetation Science*, **8**, 463–474.

Dimoudi, A. and Nikolopoulou, M. (2003) Vegetation in the urban environment: microclimatic analysis and benefits. *Energy and Buildings*, **35**, 69–76.

Dlugokencky, E. and Tans, P.P. (2015) Recent Global CO_2, NOAA, ESRS. www.esrl.noaa.gov/gmd/ccgg/trends/global.html (accessed 02/07/2015).

Doney, S.C., Fabry, V.J., Feely, R.A. *et al.* (2009) Ocean acidification: The other CO_2 problem. *Annual Review of Marine Science*, **1**, 169–192.

Driscoll, C.T., Whitall, D., Aber, J. *et al.* (2003) Nitrogen pollution in the Northeastern United States: sources, effects, and management options. *BioScience*, **53**, 357–374.

Dukes, J.S. and Mooney, H.A. (1999) Does global change increase the success of biological invaders? *Trends in Ecology and Evolution*, **14**, 135–139.

Dwyer, J.F., McPherson, E.G., Schroeder, H.W. *et al.* (1992) Assessing the benefits and costs of the urban forest. *Journal of Arboriculture*, **18**, 227–234.

Edwards, A., Russell-Smith, J., and Meyer, M. (2015) Contemporary fire regime risks to key ecological assets and processes in north Australian savannas. *International Journal of Wildland Fire*. 10.1071/WF14197

EEA. (2012) *Greenhouse Gas Emission Trends and Projections in Europe 2012 – Tracking Progress Towards Kyoto and 2020 Targets.* A Report by the European Environment Agency (EEA). Publications Office of the European Union, Luxembourg. http://www.eea.europa.eu/publications/ghg-trends-and-projections-2012 (accessed 11/12/2012)

Falkowski, P., Scholes, R.J., Boyle, E. *et al.* (2000) The global carbon cycle: A test of our knowledge of Earth as a system. *Science,* **290**, 291–296.

Fernandez, J.E. (2007) Resource consumption of new urban construction in China. *Journal of Industrial Ecology,* **11**, 99–115.

Flanagan, P.W. and Van Cleve, K. (1983) Nutrient cycling in relation to decomposition and organic-matter quality in taiga ecosystems. *Canadian Journal of Forest Research,* **13**, 795–817.

Fletcher, T.D., Mitchell, V.G., Deletic, A. *et al.* (2007) Is stormwater harvesting beneficial to urban waterway environmental flows? *Water Science & Technology,* **55**, 265–272.

Fletcher, T.D., Shuster, W., Hunt, W.F. *et al.* (2014) SUDS, LID, BMPs, WSUD, and more – The evolution and application of terminology surrounding urban drainage. *Urban Water Journal.* doi:10.1080/1573062X.2014.916314

Fowler, D., Coyle, M., Skiba, U. *et al.* (2013) The global nitrogen cycle in the twenty-first century. *Philosophical Transactions of the Royal Society of London B,* **368**, 1621. doi: 10.1098/rstb.2013.0165

Galloway, J.N., Denterner, F.J., Capone, D.G. *et al.* (2004) Nitrogen cycles: past, present, and future. *Biogeochemistry,* **70**, 153–226.

George, K., Ziska, L.H., Bunce, J.A. *et al.* (2007) Elevated atmospheric CO_2 concentration and temperature across an urban-rural transect. *Atmospheric Environment,* **41**, 7654–7665.

Golubiewski, N.E. (2003) *Carbon in Conurbations: Afforestation and Carbon Storage as Consequences of Urban Sprawl in Colorado's Front Range.* PhD Thesis, University of Colorado at Boulder.

Grimm, N.B., Faeth, S.H., Golubiewski, N.E. *et al.* (2008a) Global change and the ecology of cities. *Science,* **319**, 756–760.

Grimm, N.B., Foster, D., Groffman, P. *et al.* (2008b) The changing landscape: ecosystem responses to urbanization and pollution across climate and societal gradients. *Frontiers in Ecology and the Environment,* **6**, 264–272.

Grimm, N.B. and Redman, C.L. (2004) Approaches to the study of urban ecosystems: the case of Central Arizona-Phoenix. *Urban Ecosystems,* **7**, 199–213.

Groffman, P.M., Law, N.L., Belt, K.T. *et al.* (2004) Nitrogen fluxes and retention in urban watershed ecosystems. *Ecosystems,* **7**, 393–403.

Gruber, N. and Galloway, J.N. (2008) An Earth-system perspective of the global nitrogen cycle. *Nature,* **451**, 293–296.

Grubler, A. (1994) Technology, in *Changes in Land Use and Land Cover: A Global Perspective* (eds B.M. William and B.L. Turner), Cambridge University Press, Cambridge, pp. 287–328.

Gu, B., Ge, Y., Ren, Y. *et al.* (2012) Atmospheric reactive nitrogen in China: sources, recent trends, and damage costs. *Environmental Science and Technology,* **46**, 9420–9427.

Gu, B., Leach, A.M., Ma, L. *et al.* (2013) Nitrogen footprint in China: food, energy, and nonfood goods. *Environmental Science and Technology,* **47**, 9217–9224.

Hammond, G.P. and Jones, C.I. (2008) Embodied energy and carbon in construction materials. *Proceedings of the Institute of Civil Engineers – Energy,* **161**, 87–98.

Handia, L., Tembo, J., and Mwiinda, C. (2004) Applicability of rainwater harvesting in urban Zambia. *Journal of Science and Technology,* **1**, 1–8.

Hansell, D.A., Carlson, C.A., Repeta, D.J. *et al.* (2009) Dissolved organic matter in the ocean: a controversy stimulates new insights. *Oceanography,* **22**, 202–211.

Hatt, B.E., Fletcher, T.D., Walsh, C.J. *et al.* (2004) The influence of urban density and drainage infrastructure on the concentration and loads of pollutants in small streams. *Environmental Management*, **34**, 1112–124.

Hill, S. (2013) *Reforms for a Cleaner, Healthier Environment in China*. OECD Economics Department Working Papers 1045, OECD Publishing. Paris. 10.1787/5k480c2dh6kf-en (accessed 02/07/2015)

Hilten, R.N., Lawrence, T.M., and Tollner, E.W. (2008) Modeling stormwater runoff from green roofs with HYDRUS-1D. *Journal of Hydrology*, **358**, 288–293.

Hoegh-Guldberg, O. and Bruno, J.F. (2010) The impact of climate change on the world's marine ecosystems. *Science*, **328**, 1523–1528.

Hoornweg, D., Sugar, L., and Gomez, C.K.T. (2011) Cities and greenhouse gas emissions: moving forward. *Environment and Urbanization*, **23**, 207–227.

Hsieh, C., Davis, A.P., and Needelman, B.A. (2007) Nitrogen removal from urban stormwater runoff through layered bioretention columns. *Water Environment Research*, **79**, 2404–2411.

Idso, C.D., Idso, S.B., and Balling, R.C. Jr., (1998) The urban CO_2 dome of Phoenix, Arizona. *Physical Geography*, **19**, 95–108.

Idso, C.D., Idso, S.B., and Balling, R.C. Jr., (2001) An intensive two-week study of an urban CO_2 dome in Phoenix, Arizona, USA. *Atmospheric Environment*, **35**, 995–1000.

Imberger, S.J., Thompson, R.M., and Grace, M.R. (2011) Urban catchment hydrology overwhelms reach scale effects of riparian vegetation on organic matter dynamics. *Freshwater Biology*, **56**, 1370–1389.

Imhoff, M.L., Bounoua, L., Defries, R. *et al.* (2004) The consequences of urban land transformation on net primary productivity in the United States. *Remote Sensing of Environment*, **89**, 434–443.

IPCC. 2014. *Climate Change 2014: Synthesis Report. Contribution of Working Groups I, II and III to the Fifth Assessment Report of the Intergovernmental Panel on Climate Change* (Core Writing Team, R.K. Pachauri and L.A. Meyer (eds)). IPCC, Geneva.

Jagals, P. (1996) An evaluation of sorbitol-fermenting bifidobacteria as specific indicators of human faecal pollution of environmental water. *South African Water Research Commission*, **22**, 235.

Jo, H. and McPherson, G. (1995) Carbon storage and flux in urban residential greenspace. *Journal of Environmental Management*, **45**, 109–133.

Joos, F., Roth, R., Fuglestvedt, J.S. *et al.* (2013) Carbon dioxide and climate impulse response functions for the computation of greenhouse gas metrics: a multi-model analysis. *Atmospheric Chemistry and Physics*, **13**, 2793–2825.

Kajikawa, A., Inoue, T., and Goh, T.N. (2011) Analysis of building environment assessment frameworks and their implications for sustainability indicators. *Sustainability Science*, **6**, 233–246.

Katsiri, A. (2009) Access to improved sanitation and wastewater treatment. *ENHIS-European Environmental and Health Information Systems*. WHO Regional Office for Europe, Copenhagen. http://www.euro.who.int/__data/assets/pdf_file/0009/96957/1.3.-Access-to-improved-sanitation-and-wastewater-treatment-EDITED_layouted.pdf?ua=1 (accessed 02/07/2015)

Kaye, J.P., Groffman, P.M., Grimm, N.B. *et al.* (2006) A distinct urban biogeochemistry? *Trends in Ecology and Evolution*, **21**, 192–199.

Keith, D.A., Rodriguez, J.P., Rodriguez-Clark, K.M. *et al.* (2013) Scientific foundations for an IUCN Red List of ecosystems. *PLoS ONE* **8**, e62111.

Kennedy, C., Cuddihy, J., and Engel-Yan, J. (2007) The changing metabolism of cities. *Journal of Industrial Ecology*, **11**, 43–59.

Kennedy, C., Steinberger, J., Gasson, B. *et al.* (2009) Greenhouse gas emissions from global cities. *Environmental Science and Technology*, **43**, 7297–7302.

Kissinger, M. (2012) International trade related food miles – the case of Canada. *Food Policy*, **37**, 171–178.

Kondoh, A. and Nishiyama, J. (2000) Changes in hydrological cycle due to urbanization in the suburb of Tokyo Metropolitan Area, Japan. *Advances in Space Research*, **26**, 1173–1176.

Koont, S. (2011) *Sustainable Urban Agriculture in Cuba*, University Press of Florida, Florida.

Lee, J.H., Bang, K.W., Ketchum, L.H. *et al.* (2002) First flush analysis of urban storm runoff. *The Science of the Total Environment*, **293**, 163–175.

Leopold, L.B. (1968) *Hydrology for Urban Land Planning: a Guidebook on the Hydrological Effects of Urban Land Use*. US Geological Survey Circular 554. USGS, Washington DC. http://pubs.usgs.gov/circ/1968/0554/report.pdf (accessed 02/07/2015)

Le Quéré, C., Moriarty, R., Andrew, R.M. *et al.* (2015) Global Carbon Budget 2014. *Earth System Science Data*. doi:10.5194/essd-7-47-2015 http://www.earth-syst-sci-data.net/7/47/2015/essd-7-47-2015.html (accessed 02/07/2015)

Li, L., Yin, C., He, Q. *et al.* (2007) First flush of storm runoff pollution from an urban catchment in China. *Journal of Environmental Sciences*, **19**, 295–299.

Likens, G.E. (1992) *The Ecosystem Approach: Its Use and Abuse*. Excellence in Ecology, Vol. 3. Ecology Institute, Oldendorf/Luhe.

Lindh, G. (1972) Urbanization: a hydrological headache. *Ambio*, **1**, 185–201.

Loreau, M. (2010) Linking biodiversity and ecosystems: towards a unifying ecological theory. *Philosophical Transactions of the Royal Society B*, **365**, 49–60.

Lovell, S.T. (2010) Multifunctional urban agriculture for sustainable land use planning in the United States. *Sustainability*, **2**, 2499–2522.

Lu, C. and Tian, H. (2007) Spatial and temporal patterns of nitrogen deposition in China: Synthesis of observational data. *Journal of Geophysical Research* **112**, D22S05. doi:10.1029/2006JD007990.

Luck, G.W., Daily, G.C., and Ehrlich, P.R. (2003) Population diversity and ecosystem services. *Trends in Ecology and Evolution*, **18**, 331–336.

Mahasenan, N., Smith, S., and Humphreys, K. (2003) The cement industry and global climate change: current and potential future cement industry CO_2 emissions. in J. Gale and Y. Kaya (eds), *Greenhouse Gas Control Technologies - 6th International Conference*. Pergamon, Oxford, pp. 995–1000. 10.1016/B978-008044276-1/50157-4 (accessed 02/07/2015)

Manzoni, S., Trofymow, J.A., Jackson, R.B. *et al.* (2010) Stoichiometric controls on carbon, nitrogen, and phosphorus dynamics in decomposing litter. *Ecological Monographs*, **80**, 89–106.

Maurin, F.D. (2011) Fukushima: Consequences of systemic problems in nuclear plant design. *Economic and Political Weekly*, **46**, 10–12.

McClintock, N. (2010) Why farm a city? Theorizing urban agriculture through a lens of metabolic rift. *Cambridge Journal of Regions, Economy and Society*, **3**, 191–207.

McPherson, G., Simpson, J.R., Peper, P.J. *et al.* (2005) Municipal forest benefits and costs in five US cities. *Journal of Forestry*, **103**, 411–416.

Mentens, J., Raes, D., and Hermy, M. (2006) Green roofs as a tool for solving the rainwater runoff problem in the urbanized 21st century? *Landscape and Urban Planning*, **77**, 217–226.

Millennium Ecosystem Assessment (2003) *Ecosystems and Human Well-Being: A Framework for Assessment*, Island Press, Washington DC.

Mok, H., Williamson, V.G., Grove, J.R. *et al.* (2014) Strawberry fields forever? Urban agriculture in developed countries: a review. *Agronomy for Sustainable Development*, **34**, 21–43.

Moomaw, W.R. (2002) Energy, industry and nitrogen: strategies for decreasing reactive nitrogen emissions. *Ambio*, **31**, 184–189.

Moomaw, W.R., Burgherr, P., Heath, G. *et al.* (2011) Annex II: Methodology, in *IPCC Special Report on Renewable Energy Sources and Climate Change Mitigation* (eds O. Edenhofer, R. Pich-Madruga, Y. Sokona, *et al.*), Cambridge University Press, Cambridge, pp. 973–1000.

Mougeot, L.J.A. (ed) (2005) *AGROPOLIS: The Social, Political, and Environmental Dimensions of Urban Agriculture*, Earthscan, London.

Mu, Q., Zhao, M., and Running, S.W. (2011) Improvements to a MODIS global terrestrial evapotranspiration algorithm. *Remote Sensing of Environment*, **115**, 1781–1800.

Norton, B.A., Bosomworth, K., Coutts, A. *et al.* (2013) *Planning for a Cooler Future: Green Infrastructure to Reduce Urban Heat.* Victorian Centre for Climate Change Adaptation Research, Melbourne. http://www.vcccar.org.au/sites/default/files/publications/VCCCAR Green Infrastructure Guide Final.pdf (accessed 25/06/2015)

Norton, B.A., Coutts, A.M., Livesley, S.J. *et al.* (2015) Planning for cooler cities: A framework to prioritise green infrastructure to mitigate high temperatures in urban landscapes. *Landscape and Urban Planning*, **134**, 127–138.

Nowak, D.J. and Crane, D.E. (2002) Carbon storage and sequestration by urban trees in the USA. *Environmental Pollution*, **116**, 381–389.

Nowak, D.J., Crane, D.E., and Stevens, J.C. (2006) Air pollution removal by urban trees and shrubs in the United States. *Urban Forestry and Urban Greening*, **4**, 115–123.

Oberndorfer, E., Lundholm, J., Bass, B. *et al.* (2007) Green roofs as urban ecosystems: ecological structures, functions, and services. *BioScience*, **57**, 823–833.

O'Brien, D. and Hes, D. (2008) The third way: developing low environmental impact housing prototypes for hot/humid climates. *International Journal for Housing Science*, **32**, 311–322.

Odum, E.P. (1983) *Basic Ecology*, Saunders Publishing, Philadelphia.

Okoko, E. (2008) The urban storm water crisis and the way out: empirical evidences from Ondo Town, Nigeria. *The Social Sciences*, **3**, 148–156.

Olivier, J.G.J., Janssens-Maenhout, G., Peters, J.A.H.W. *et al.* (2011) *Long-term Trend in Global $C0_2$ Emissions. 2011 Report.* PBL Netherlands Environmental Assessment Agency and the European Union, The Hague. http://www.pbl.nl/sites/default/files/cms/publicaties/C02 Mondiaal_ webdef_19sept.pdf (accessed 03/07/2015)

Owusu-Asante, Y. and Ndiritu, J. (2009) The simple modelling method for storm- and grey-water quality management applied to Alexandra settlement. *Water SA*, **35**, 615–626.

Owusu-Asante, Y. and Stephenson, D. (2006) Estimation of storm runoff loads based on rainfall-related variables and power law models – case study in Alexandra. *Water SA*, **32**, 1–8.

Parkinson, J. (2003) Drainage and stormwater management strategies for low-income urban communities. *Environment and Urbanization*, **15**, 115–126.

Parkinson, J., Tayler, K., and Mark, O. (2007) Planning and design of urban drainage systems in informal settlements in developing countries. *Urban Water Journal*, **4**, 137–149.

Parris, K.M. and Hazell, D.L. (2005) Biotic effects of climate change in urban environments: the case of the grey-headed flying-fox (*Pteropus poliocephalus*) in Melbourne. *Australia. Biological Conservation.*, **124**, 267–276.

Pataki, D.E., Alig, R.J., Fung, A.S. *et al.* (2006) Urban ecosystems and the North American carbon cycle. *Global Change Biology*, **12**, 2092–2102.

Pauchard, A., Aguayo, M., Pena, E. *et al.* (2006) Multiple effects of urbanization on the biodiversity of developing countries: the case of a fast-growing metropolitan area (Concepción, Chile). *Biological Conservation*, **127**, 272–281.

Paul, M.J. and Meyer, J.L. (2001) Streams in the urban landscape. *Annual Review of Ecology and Systematics*, **32**, 333–365.

Pickett, S.T.A., Cadenasso, M.L., Grove, J.M. *et al.* (2001) Urban ecological systems: linking terrestrial ecological, physical, and socioeconomic components of metropolitan areas. *Annual Review of Ecology, Evolution, and Systematics*, **32**, 127–157.

Pickett, S.T.A. and Grove, J.M. (2009) Urban ecosystems: what would Tansley do? *Urban Ecosystems*, **12**, 1–8.

Pidwirny, M. 2008. *Fundamentals of Physical Geography*, 2nd edn (online text book). PhysicalGeography.net

Pouyat, R.V., Russell-Anelli, J., Yesilonis, I.D. *et al.* (2003) *Soil Carbon in Urban Forest Ecosystems*, CRC Press, Boca Raton.

Prather, M.J., Holmes, C.D. and Hsu, J. (2012) Reactive greenhouse gas scenarios: systematic exploration of uncertainties and the role of atmospheric chemistry. *Geophysical Research Letters* **39**, L09803. doi: 10.1029/2012GL051440

Prentice, I.C., Farquhar, G.D., Fasham, M.J.R. *et al.* (2001) The carbon cycle and atmospheric carbon dioxide, in *Climate Change 2001: The Scientific Basis* (eds J.T. Houghton, Y. Ding, D.Y. Griggs, *et al.*), Cambridge University Press, Cambridge, pp. 183–237.

Rabalais, N.N. (2002) Nitrogen in aquatic ecosystems. *AMBIO: A Journal of the Human Environment*, **31**, 102–112.

Reddy, B.V.V. and Jagadish, K.S. (2003) Embodied energy of common and alternative building materials and technologies. *Energy and Buildings*, **35**, 129–137.

Reich, P.B. and Hobbie, S.E. (2013) Decade-long soil nitrogen constraints on the CO_2 fertilization of plant biomass. *Nature Climate Change*, **3**, 278–282.

Roth, M. (2007) Review of urban climate research in (sub)tropical regions. *International Journal of Climatology*, **27**, 1859–1873.

Seitzinger, S.P. and Kroeze, C. (1998) Global distribution of nitrous oxide production and N inputs in freshwater and coastal marine ecosystems. *Global Biogeochemical Cycles*, **12**, 93–113.

Smit, J. and Nasr, J. (1992) Urban agriculture for sustainable cities: using wastes and idle land and water bodies as resources. *Environment and Urbanization*, **4**, 141–152.

Stoffberg, G.H., van Rooyen, M.W., van der Linde, M.J. *et al.* (2010) Carbon sequestration estimates of indigenous street trees in the city of Tshwane, South Africa. *Urban Forestry and Urban Greening*, **9**, 9–14.

Stone, B. Jr., (2004) Paving over paradise: how land use regulations promote residential imperviousness. *Landscape and Urban Planning*, **69**, 101–113.

Suppiah, R. and Whetton, P.H. (2007) *Projected Changes in Temperature and Heating Degree-Days for Melbourne and Victoria, 2008-2012*. CSIRO Marine and Atmospheric Research, Aspendale. http://www.ccma.vic.gov.au/soilhealth/climate_change_literature_review/documents/organisations/csiro/MelbourneEDD2008_2012.pdf (accessed 02/07/2015)

Svirejeva-Hopkins, A., Schellnhuber, H.J., and Pomaz, V.L. (2004) Urbanised territories as a specific component of the global carbon cycle. *Ecological Modelling*, **173**, 295–312.

Tansley, A.G. (1935) The use and abuse of vegetation concepts and terms. *Ecology*, **16**, 284–307.

Tarnocai, C., Canadell, J.G., Schuur, E.A.G. *et al.* (2009) Soil organic carbon pools in the northern circumpolar permafrost region. *Global Biogeochemical Cycles* **23**, GB2023. DOI: 10.1029/2008GB003327

Tarvainen, O., Markkola, A.M., and Strommer, R. (2003) Diversity of macrofungi and plants in Scots pine forests along an urban pollution gradient. *Basic and Applied Ecology*, **4**, 547–556.

Taylor, S.L., Roberts, S.C., Walsh, C.J. *et al.* (2004) Catchment urbanization and increased benthic algal biomass in streams: linking mechanisms to management. *Freshwater Biology*, **49**, 835–851.

Tscharntke, T., Klein, A.M., Steffan-Dewenter, I. *et al.* (2005) Landscape perspectives on agriculture intensification and biodiversity – ecosystem service management. *Ecology Letters*, **8**, 857–874.

Turner, M.G., Collins, S.L., Lugo, A.L. *et al.* (2003) Disturbance dynamics and ecological response: the contribution of long-term ecological research. *BioScience*, **53**, 46–56.

Tyrvainen, L., Pauleit, S., Seeland, K. *et al.* (2005) Benefits and uses of urban forests and trees, in *Urban Forests and Trees* (eds C.C. Konijnendijk, K. Nilsson, T.B. Randrup, and J. Schipperijn), Springer, Berlin Heidelberg, pp. 81–114.

United Nations. 1998. *Kyoto Protocol to the United Nations Framework Convention on Climate Change*. United Nations Framework Convention on Climate Change, Bonn. http://unfccc.int/resource/docs/convkp/kpeng.pdf (accessed 02/07/2015)

United Nations. 2012. *Doha Amendment to the Kyoto Protocol*. United Nations Framework Convention on Climate Change, Bonn. http://unfccc.int/files/kyoto_protocol/application/pdf/kp_doha_amendment_english.pdf (accessed 02/07/2015)

United Nations. 2014. *World Urbanization Prospects: The 2014 Revision, CD-ROM Edition*. United Nations Department of Economic and Social Affairs, Population Division, New York. http://esa.un.org/unpd/wup/ (accessed 08/06/2015)

United Nations. (2015) Adoption of the Paris Agreement. United Nations Framework Convention on Climate Change, Bonn. http://unfccc.int/resource/docs/2015/cop21/eng/l09r01.pdf (accessed 21/01/2016)

van Aardenne, J.A., Dentener, F.J., Olivier, J.G.J. *et al.* (2001) A 1×1 resolution data set of historical anthropogenic trace gas emissions for the period 1890-1990. *Global Biogeochemical Cycles*, **15**, 909–928.

van Veenhuizen, R. (ed). (2006) *Cities Farming for the Future: Urban Agriculture for Green and Productive Cities*. RUAF Foundation, IDRC and IIRR, Silang.

Vitousek, P.M., Aber, J.D., Howarth, R.W. *et al.* (1997) Human alteration of the global nitrogen cycle: Sources and consequences. *Ecological Applications*, **7**, 737–750.

Walsh, C.J., Fletcher, T.D., and Burns, M.J. (2012) Urban stormwater runoff: a new class of environmental flow problem. *PLoS ONE*, **7**, 1–10.

Walsh, C.J., Roy, A.H., Feminella, J.W. *et al.* (2005) The urban stream syndrome: current knowledge and the search for a cure. *Journal of North American Benthological Society*, **24**, 706–723.

Wang, D., Heckathorn, S.A., Wang, X. *et al.* (2012) A meta-analysis of plant physiological and growth responses to temperature and elevated CO_2. *Oecologia*, **169**, 1–13.

White, P.S. and Pickett, S.T.A. (1985) The ecology of natural disturbance and patch dynamics, in *The Ecology of Natural Disturbance and Patch Dynamics* (eds S.T.A. Pickett and P.S. White), Academic Press, New York, pp. 3–13.

Willis, A.J. (1997) *Forum. Functional Ecology*, **11**, 268–271.

Wong, T.H.F. (2007) Water sensitive urban design: the journey thus far. *Australian Journal of Water Resources*, **110**, 213–222.

World Steel Association. (2013) *Steel's Contribution to a Low Carbon Future: World Steel position paper*. World Steel Association, Brussels. http://www.worldsteel.org/publications/ position–papers/Steels-contribution-to-a-low-carbon-future.html (accessed 06/06/2015).

World Steel Association. (2014) *Crude steel production 2014-2015*. World Steel Association, Brussels. http://www.worldsteel.org/statistics/crude-steel-production.html (accessed 02/06/2015).

Yudelson, J. (2007) *Green Building A to Z: Understanding the Language of Green Building*, New Society Publishers, Gabriola Island.

Zehr, J.P. and Kudela, R.M. (2011) Nitrogen cycle of the open ocean: from genes to ecosystems. *Annual Review of Marine Science*, **3**, 197–225.

CHAPTER 6

The urban ecology of humans

6.1 Introduction

In this era of accelerating urbanization, more people live in cities than ever before – both in absolute numbers and as a proportion of the total human population. At their best, cities are hubs of human endeavour, creativity, community and purposeful living, offering a critical mass of people that brings a sense of place, a sense of diversity and a sense of belonging. They offer opportunity, employment, engagement, fulfilment and hope for a better future (Hes and Du Plessis 2015). In general, rates of poverty are lower in urban than rural areas (World Bank 2013; Figure 6.1). Well-designed cities provide a space- and energy-efficient way to meet the economic and social needs of humans on a large scale (Kennedy et al. 2007; Kennedy et al. 2011; Farr 2012). The most prosperous residents of cities enjoy a high standard of living and amenity, with secure homes, electricity, clean water, adequate ventilation and effective heating and cooling, as well as access to quality education and health care. Affluent urban neighbourhoods tend to be neat and ordered with abundant parks and street trees offering amenity, shade and a cooler microclimate in summer (Heynen 2003; Harlan et al. 2006; Zhu and Zhang 2008; Huang et al. 2011b; Figure 6.2a).

However, at their worst, cities can be hubs of unemployment, violence, poverty, grime and human despair. The vastness of cities, the density of their human population and the greyness of the built form in concrete and asphalt can make cities impersonal and alienating. The poorest residents of many cities live in substandard, overcrowded, insecure housing, often without effective heating and cooling, adequate ventilation and sanitation or access to clean drinking water (Weeks et al. 2007; Montgomery 2009). In most cities, poorer neighbourhoods have a lower cover of vegetation, fewer parks and street trees, and a higher cover of impermeable surfaces (Pedlowski et al. 2002; Heynen 2003; Escobedo 2006; Harlan et al. 2006; Weeks et al. 2007; Huang et al 2011b; Shanahan et al. 2011; Clarke et al. 2013; Figure 6.2b). For example, the cover of vegetation in the Nima and Maamboi slums of Accra, Ghana was estimated to be 6%, with an impervious-surface cover of 84% (due mainly to the roofs of densely-crowded dwellings) and a bare-soil cover of 10% (Weeks et al. 2007). Even in planned

Ecology of Urban Environments, First Edition. Kirsten M. Parris.

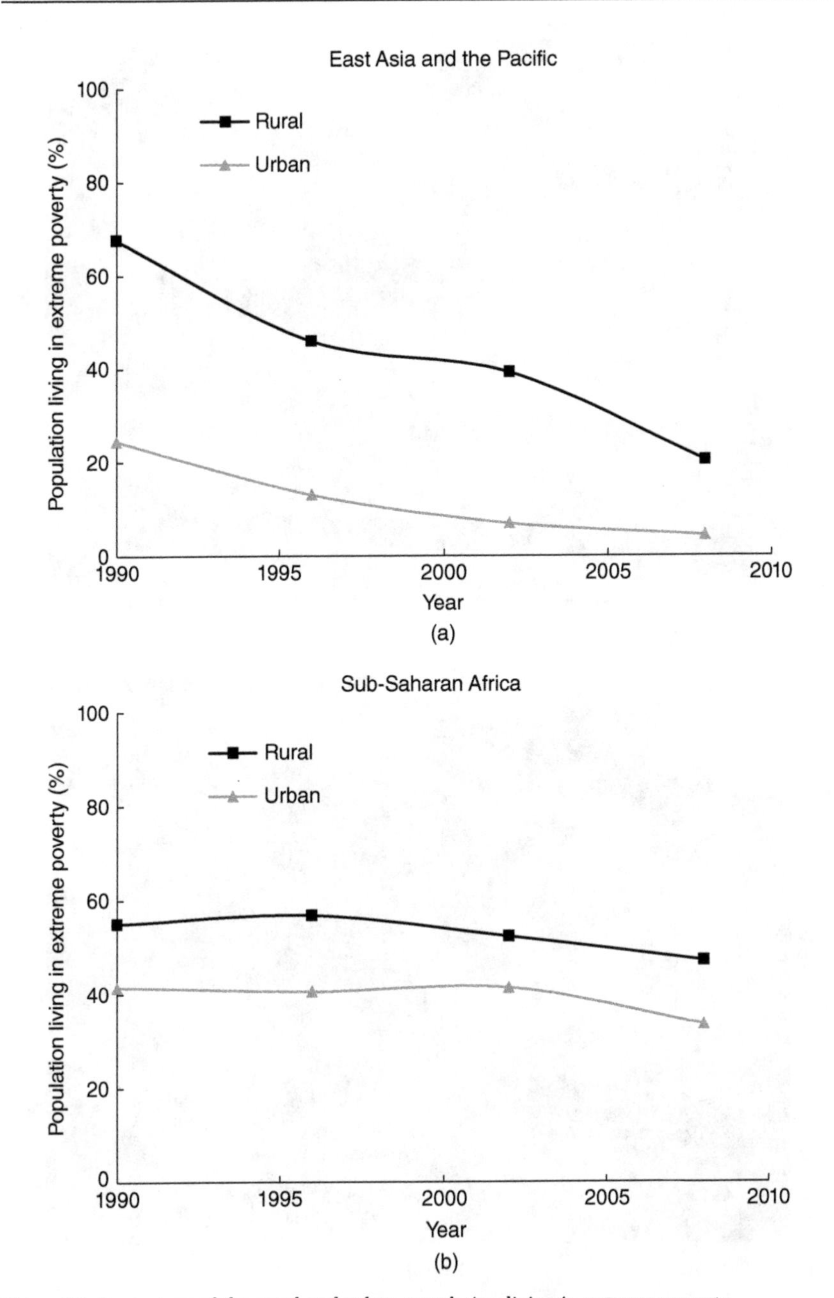

Figure 6.1 Percentage of the rural and urban population living in extreme poverty, 1990–2008, in two regions of the world: (a) East Asia and the Pacific and (b) Sub-Saharan Africa. Extreme poverty is defined as living below US$1.25/day. Data from World Bank 2013.

(a)

(b)

Figure 6.2 Contrasting urban green cover: (a) the Atlanta Botanical Gardens in Atlanta, Georgia, USA. Atlanta is known as a city of trees, supporting 48% tree cover within the city limits (Giarrusso et al. 2014). (Picture has been cropped and converted to black and white.) Photograph by Eric Sonstroem. Used under CC-BY-2.0 https://creativecommons.org/licenses/by/2.0/. (b) An informal settlement (favela) in Brazil. Photograph by Unsplash. https://pixabay.com/en/city-urban-slum-favela-buildings-731385/. Used under CC0 1.0 https://creativecommons.org/publicdomain/zero/1.0/deed.en

(non-slum) areas of cities, vegetation cover varies with the socioeconomic status of the neighbourhood; for example, from an average of 33% tree cover in affluent municipalities or *comunas* in Santiago, Chile to an average of 12% in poorer municipalities (Escobedo et al. 2006).

How can we best frame a consideration of the urban ecology of humans? Each of the disciplines involved in the study of people and society in cities – including sociology, human geography, urban planning, demography, psychology, criminology, epidemiology and public health – has its own theoretical basis and methodological approach. The discipline of ecology can also be applied to humans; human ecology is the study of relationships between humans and their natural, social and constructed environments. It is often conceptualized as an interdisciplinary field encompassing geography, anthropology, psychology, biology, sociology and urban planning (The Editors 1972; Bates 2012). These disciplines are all important for the improved understanding and management of cities. However, in this chapter I consider how the processes of urbanization and the experience of living in an urban environment affect humans as individuals and populations from the perspective of a single discipline – the biological discipline of ecology. Characteristics of the urban environment, from the nature of the urban form to air pollution, heat, noise, access to clean drinking water and sanitation, have both sub-lethal and lethal effects on individual humans (Schell and Denham 2003; Miller et al. 2007; Valavanidis et al. 2008; Montgomery 2009; Hammer et al. 2014). Scaling up from individuals to populations, these characteristics also influence the vital rates of human populations (births, deaths, immigration and emigration) in cities around the world.

Central to this analysis is the spatial diversity of the urban human experience, both within a single city and between different cities. The urban human experience can also vary in the same location over time, as cities follow various trajectories from their early foundations to expansion and industrialization, followed by recession, urban decay and, in some cases, further urban renewal (Wyly and Hammel 1999; Martinez-Fernandez et al. 2012; Zheng et al. 2014). One such trajectory is captured in the lyrics of *Telegraph Road* (see Preface), which describes the transformation of a thoroughfare in Detroit, Michigan from dirt track to multi-lane road; the city around it from a single cabin to an extensive, industrialized metropolis; and the experience of its residents from the optimism of early settlement, expansion and economic boom times, to the despair of unemployment, urban decay and alienation (Knopfler 1982). The current, rapid expansion of cities in Africa, Asia and South America is largely a consequence of mass immigration from impoverished rural areas, and in many cases is characterized by the construction of informal settlements such as slums and shanty towns (UN-Habitat 2003; Cilliers 2008; WHO & UN-Habitat 2010; Shin and Li 2012). The urban experiences of these new migrants will be quite unlike those of long-time residents who live in planned and relatively well-resourced parts of the city.

6.2 The urban form

6.2.1 Urban parks and open space

As explored in earlier chapters, the construction of cities involves a variety of processes that combine to change a landscape of green and brown, of soft surfaces and open space, of clean air and water, of natural sounds and nighttime darkness, to one of grey, hard surfaces, of densely constructed buildings and roads, of polluted air and water, and of anthropogenic noise and nocturnal light. These characteristics of cities impact the health and wellbeing of the people who live there, acting to separate them from nature and in some cases also to separate them from each other (Silver 1997; Turner et al. 2004; Miller 2005; Soule 2009; Ross 2011; but see Box 6.1). Many city dwellers live in neighbourhoods that are biologically depauperate, and therefore only have the opportunity to interact with a much-reduced diversity of non-human species.

Box 6.1 Human–Wildlife Interactions in Cities

Interactions with wildlife (i.e., non-domesticated animals) in private gardens, streetscapes or public parks and reserves give city residents additional opportunities to connect with nature in the urban landscape. Interactions between humans and wildlife can be classified along a gradient of intensity, from simple observation of invertebrates, birds, reptiles, frogs and other animals to provision of food and shelter, competition for resources, damage to property, transmission of disease, and injury or death (of both humans and wildlife). Positive interactions between humans and wildlife in cities can be mutually beneficial. For example, many people enjoy bird watching in their own gardens or the wider neighbourhood, while bird baths and bird feeders can provide important resources for birds (Miller 2005; Jones and Reynolds 2008; Figure 6.3). The feeding of garden birds can be considered as a physical manifestation of a continuing connection between people and nature despite the highly modified character of urban environments (Fuller at al. 2008), and this connection is important for human wellbeing (Maller and Townsend 2004; Miller 2005; Jones and Reynolds 2008).

Supplementary feeding of birds increases the diversity and abundance of birds visiting individual gardens, as well as the total abundances of birds across neighbourhoods (Chamberlain et al. 2005; Daniels and Kirkpatrick 2006; Parsons et al. 2006; Fuller et al. 2008). However, provision of large quantities of certain food types (e.g., an estimated 60,000 tonnes/year of seed and peanuts in the UK; Glue 2006) may favour granivorous birds over other guilds such as insectivores and nectarivores, which can change the composition of assemblages (Allen and O'Connor 2000). In addition, feeding bread to ducks and other water fowl may increase intra- and inter-specific aggression in urban bird communities and increase dependence on a food source of poor nutritional quality (Campbell 2008; Chapman and Jones 2009; Chapman and Jones 2011).

Direct and indirect conflict between humans and wildlife in urban environments is increasing in many parts of the world as cities expand into previously wild habitats. Conflicts arise when animals exploit particular urban food resources against the wishes of humans, such as flowers, fruit and vegetables from private gardens; food scraps, litter and garbage; and food provided by humans for pets or another group of wild animals such as birds (Banks et al. 2003; Beckmann and Berger 2003; Harper et al. 2008; Lamarque et al. 2009; Adams et al. 2013). Certain medium-sized predators such as the red fox, raccoon and Eurasian badger are well adapted to city life, and often maintain higher population densities in urban and suburban areas than in rural areas (Bateman and Fleming 2012; Šálek et al. 2015). In Melbourne, Australia, the introduced red fox commonly kills domestic chickens along with native wildlife such as the endangered southern brown bandicoot, swamp skink and growling grass frog (Saunders et al. 1995; Green 2014).

Figure 6.3 An eastern Rosella *Platycercus eximius* on a back-yard bird feeder. (Picture has been cropped and converted to black and white.) Photograph by Lesley Smitheringale (originally posted to Flickr as Supper Time) https://commons.wikimedia.org/wiki/File %3AEastern_Rosella_(Platycercus_eximius)4_-bird_feeder.jpg. Used under CC BY 2.0 http://creativecommons.org/licenses/by/2.0.

Larger carnivores including cougars, leopards and lions may prey on household pets such as cats and dogs, and in some instances, on people themselves (McKee 2003; Thornton and Quinn 2009; Bhatia et al. 2013). In North America, coyotes and cougars sometimes attack domestic animals and humans (Beier 1991; Thornton and Quinn 2009; White and Gehrt 2009). The incidence of coyote attacks in urban and suburban environments of the USA and Canada appears to be increasing as coyote populations grow (White and Gehrt 2009). In Mumbai, India, there were 101 confirmed leopard attacks on humans resulting in injury or death between 2000 and 2010, with the bulk of attacks occurring between 2001 and 2005 (Bhatia et al. 2013). The leopards belonged to a population from Sanjay Gandhi National Park, which is situated next to residential areas of Mumbai that support a human population density of >30,000 individuals/km^2. In many instances of human–wildlife conflict, human interests take precedence over the interests of wildlife, with "problem" animals captured and relocated or euthanized (Choudhury 2004; Spencer et al. 2007; Bhatia et al. 2013). However, relocation of some groups of animals – such as mammals and reptiles – to avert human–wildlife conflict has a low probability of success, with many relocated individuals returning to the site of capture and others suffering high mortality rates in their new environment (e.g., Pietsch 1994; Linnell et al. 1997; Sullivan et al. 2004; Germano and Bishop 2009).

For example, in a study of neighbourhood biological diversity in five cities (Tucson, Arizona; Berlin, Germany; Washington, DC; Florence, Italy; Chiba City, Japan), up to 84% of residents lived in neighbourhoods with below-average diversity of birds and ferns (Turner et al. 2004). This was due to an inverse correlation between human population density and biodiversity, with the more biodiverse parts of cities, such as natural parklands and areas of remnant vegetation, clustered around the urban fringe or in other neighbourhoods with fewer humans per unit area. Many residents were consequently separated from much of the biodiversity that cities still manage to support (Miller 2005). The wedge between humans and nature in cities can drive a phenomenon known as the extinction of experience (Pyle 1978); a cycle of experiential impoverishment that begins with the loss of local flora and fauna, followed by alienation from and indifference to nature. This, in turn, may lead to a further loss of biodiversity and a deeper isolation from the natural world (Pyle 1978; Miller 2005). It has been suggested that this isolation may reduce the commitment of humans to biological conservation more broadly (i.e., beyond the urban sphere; Miller 2005), but it also has significant implications for human health and wellbeing.

The opportunity to connect with nature in urban reserves, neighbourhood parks or even via an individual tree can have a range of positive effects on mental health and psychological wellbeing, reducing stress and mental fatigue, restoring attention, enhancing the ability to think and gain perspective, and providing a sense of attachment to place and higher life satisfaction (Kuo 2001; Grahn and Stigsdotter 2003; Chiesura 2004; Fuller et al. 2007; Chaudhry et al. 2011; Haq 2011; reviewed by Maller et al. 2004; Clark et al. 2014; Shanahan et al. 2015). For example, a meta-analysis of the mental health benefits of "green exercise" (exercising in the presence of nature) found a consistent, positive effect on self-esteem and mood across 10 studies (overall effect size: $d = 0.46$ (CI 0.34–0.59) for self-esteem and $d = 0.54$ (CI 0.38–0.69) for mood; Barton and Pretty 2010). Interestingly, dose responses showed that substantial mental-health benefits could be gained from only five minutes of green exercise, and that the positive effect on mood and self-esteem of exercising in urban green spaces was similar to that gained by exercising in some rural green spaces, such as farmland, forest and woodland. However, exercising in green nature with water views provided greater benefits for mood and self-esteem.

Public parks and communal gardens can also act as a focal point for community engagement, increasing interactions between neighbours, improving social cohesion and strengthening social bonds (Coley et al. 1997; Saldivar-Tanaka and Krasny 2004; Kingsley and Townsend 2006; Maas et al. 2009). A survey of >10,000 residents in The Netherlands found that a higher cover of green space in a 1-km and 3-km radius around a person's home was correlated with lower levels of loneliness and a reduced perception of social isolation, suggesting that greater social contact may be one mechanism underlying the observed positive relationship between green space and health (Maas et al. 2009). Simply viewing green rooftop gardens though a window can improve attention span and mood in city office workers (Lee et al. 2015), while a classic study by Ulrich (1984) found patients recover from surgery more quickly if they have a view of nature from their hospital window. Interestingly, a study in Sheffield, England observed that the sense of psychological wellbeing people gain from urban green space increases with perceived species richness, suggesting that biological complexity is a key component of humans' enjoyment of nature and hence a valuable characteristic of urban parks (Fuller et al. 2007; also see Carrus et al. 2015).

Access to open space in cities also improves the physical health of city residents by providing space and amenities for exercise and recreation (Sallis et al. 2011; Lowe et al. 2014). Particularly in affluent countries, insufficient exercise contributes to chronic diseases such as obesity, heart disease, cancer and Type 2 diabetes (Kaczynski and Henderson 2007; Bauman et al. 2012; Lowe et al. 2014). Chronic diseases have now overtaken infectious diseases (Box 6.2) as the leading cause of illness and death in urban populations (Lowe et al. 2014). Recent data from 122 countries around the world indicate that on average, 31.1% (95% CI 30.9–31.2%) of adults and 80.3% (80.1–80.5%) of adolescents do not achieve recommended levels of exercise (Hallal et al. 2012). Inactivity levels vary by region, from 17% (16.8–17.2%) in south-east Asia to >40% in the Americas and the eastern Mediterranean (Figure 6.4). The availability, proximity, accessibility and quality of parks in an urban neighbourhood are all correlated with higher physical activity among residents (Veitch et al. 2014). Parks encourage physical activity by providing a pleasant destination to which people can walk or cycle and a suitable environment in which to exercise (Pikora et al. 2003; Cervero et al. 2009; Sugiyama et al. 2010, 2012). In Perth, Australia, residents with very good access to large, attractive public open spaces are 50% more likely to walk regularly (six or more walking sessions/week, totalling ≥3 hours) than those with very poor access (Giles-Corti et al. 2005).

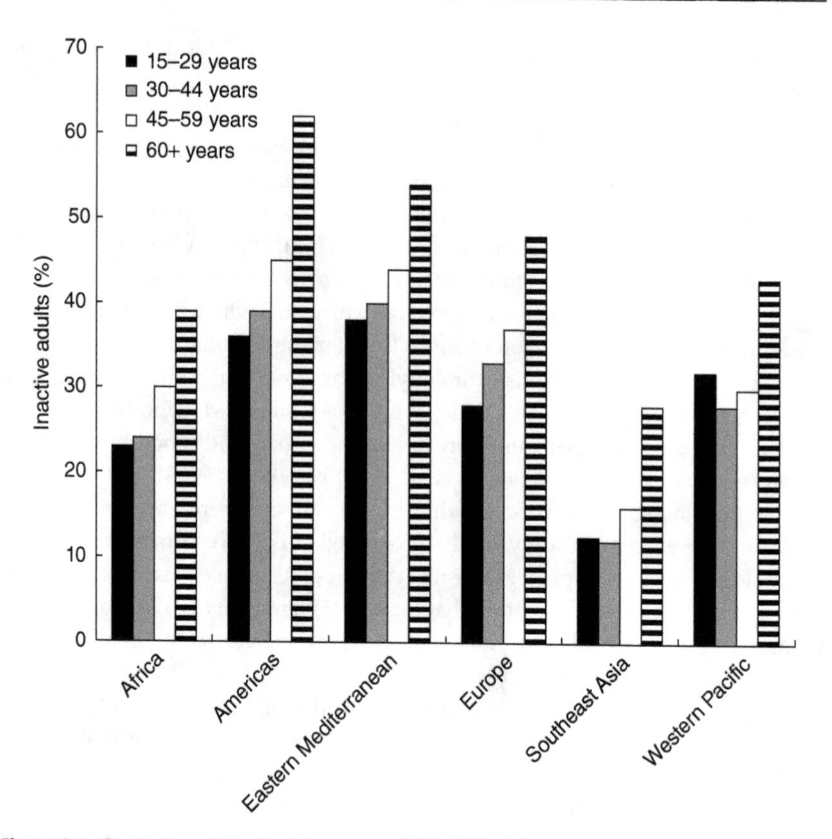

Figure 6.4 Physical inactivity by age group and World Health Organisation region (% of population); physical inactivity varies by region and tends to increase with age. Adapted from Hallal et al. 2012, Figure 2. Reproduced with permission of Elsevier.

Box 6.2 Infectious Diseases in Urban Environments

Throughout history, high population densities, malnutrition and poor sanitary conditions in cities have combined to promote the spread of infectious diseases (Redman and Jones 2004). The early cities of the world contributed to the evolution and spread of a series of infectious diseases that have been responsible for widespread morbidity and mortality over the centuries (Redman 1999). This trend continued during the industrial revolution, when overcrowded, squalid cities in Western Europe and Britain were centres for diseases such as smallpox, dysentery, tuberculosis, typhoid fever, typhus and cholera. Cholera is caused by a bacterium transmitted by ingestion of water or food contaminated by the faeces of an infected person. The disease can spread rapidly when supplies of drinking water are contaminated with sewage or untreated human waste.

Outbreaks of cholera were common in 19th century London prior to construction of the city's sewers, with the most famous outbreak occurring in Broad Street, Soho in 1854. A local physician, John Snow, mapped the pattern of cholera cases and identified a water pump on Broad Street as the source of the infection, as nearly all deaths had occurred within a short distance of this pump (Johnson 2006; Figure 6.5). He persuaded the local authorities to disable the pump, even though the germ theory of disease was not widely accepted at that time. This event is regarded as a foundational moment in the development of epidemiology. Cholera still occurs in cities, as demonstrated by the outbreak in Harare, Zimbabwe in 2008, caused by the collapse of the urban water supply and sanitation systems combined with the onset of the rainy season (WHO 2011). This outbreak spread across Zimbabwe and into a number of neighbouring countries, with an estimated death toll of >4000 (Mason 2009; Mukandavire et al. 2011). Other infectious diseases common in cities (particularly informal settlements) today include tuberculosis, dengue fever, HIV, pneumonia and diarrhoeal diseases (WHO & UN-Habitat 2010).

Figure 6.5 A variant of the original map drawn by Dr John Snow, British physician and one of the founders of medical epidemiology, showing the location of deaths from cholera in Soho, London in 1854 (black circles). The source of the infection was a water pump on Broad Street.

6.2.2 Urban sprawl and car dependence

Low urban density – also known as urban sprawl – increases dependence on private cars for transport while reducing opportunities for residents to walk or cycle around their neighbourhood and to and from work (Ewing et al. 2003; Dieleman and Wegener 2004). Sprawling residential developments located far from places of work, public transport and amenities such as shops and schools are characteristic of many cities around the world. These kinds of neighbourhoods discourage physical activity and are positively correlated with overweight and obesity among residents (Lopez 2004; Garden and Jalaludin 2009). A study of residents in Atlanta, Georgia, found that land-use mix (as calculated by the evenness of cover of four land-use types; residential, commercial, office, and institutional) within 1 km of each person's household was correlated with a reduced likelihood of obesity (BMI >30) across all gender and ethnic groups (Frank et al. 2004; Figure 6.6). Each additional hour spent in a car per day was associated with a 6% increase in the likelihood of obesity, while each kilometre walked per day decreased the likelihood of obesity by 4.8% (Frank et al. 2004).

Higher residential densities, good street connectivity (a gridded network of streets, rather than many disconnected streets and cul de sacs), access to walking

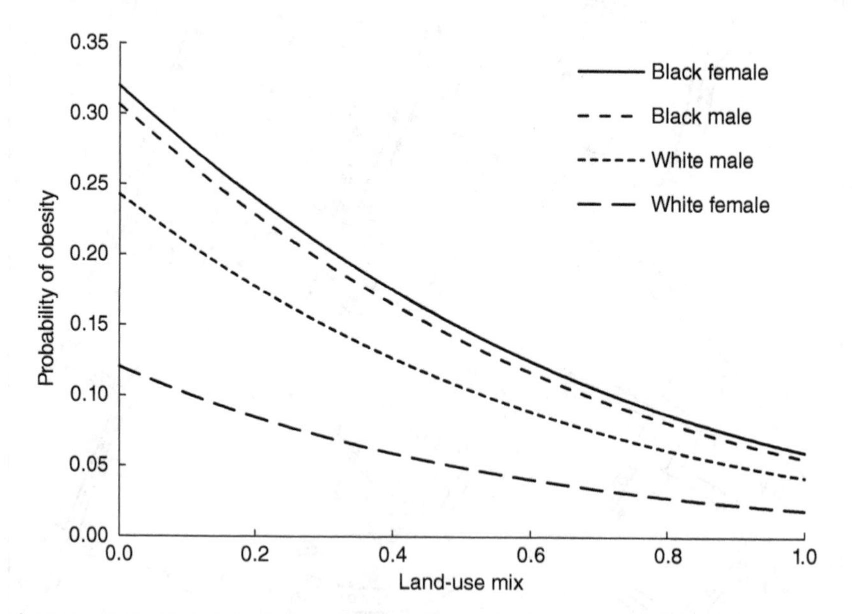

Figure 6.6 Probability of obesity as a function of urban land-use mix, ethnicity and gender in Atlanta, Georgia, USA. The probability of obesity declines across all social groups as land-use mix increases. Obesity is defined as a body-mass index ≥30. Land-use mix is the evenness of cover by area of four types of land use (residential, commercial, office and institutional) within 1 km of each participant's household. Adapted from Frank et al. 2004, Figure 2. Reproduced with permission of Elsevier.

and cycle paths, and mixed land use (with shorter distances between homes, shops and places of work) are correlated with higher levels of walking and cycling for transport (Lowe et al. 2014). The walkability of urban neighbourhoods has gained increasing research and policy attention in recent years (WAPC 2007; NSW Government 2004; Saelens et al. 2003; Giles-Corti et al. 2015), and the walkability of a neighbourhood is negatively correlated with rates of obesity amongst its residents (Smith 2008; Sallis et al. 2009; Casagrande et al. 2011). These individual-level effects can be scaled up and applied to known relationships between obesity and mortality from chronic illnesses, such as ischaemic heart disease and Type 2 diabetes, to estimate the impact of urban sprawl or neighbourhood walkability on population mortality rates (e.g., Hankey et al. 2012).

Walkable neighbourhoods can be identified by the three Ds: population *density*, pedestrian-friendly *design*, and a *diversity* of destinations (Cervero and Kockelman 1997; Smith 2008). For example, the density and connectivity of roads strongly influences rates of utilitarian (non-recreational) walking and cycling in Bogotá, Columbia (Cervero et al. 2009). Residents in neighbourhoods with medium or high connectivity are twice as likely to walk for ≥30 minutes/weekday than those in neighbourhoods with low connectivity (with the neighbourhood defined by a 500-m radius around the person's block of residence). Similarly, residents are twice as likely to cycle for ≥30 minutes/weekday for non-recreational purposes (e.g., travelling to work or school, shopping, visiting the doctor) where road density is ≥0.2 road km/km^2 of land area in a 1000-m radius of their block of residence than where road density is <0.2 road km/km^2 (Cervero et al. 2009). In many cities, older neighbourhoods are more likely to be designed for pedestrians, with shaded sidewalks, narrower streets, slower traffic and more diverse destinations; newer neighbourhoods are more likely to be designed for the efficient flow of vehicular traffic (Smith 2008).

6.2.3 Neighbourhood disadvantage and neighbourhood disorder

Characteristics of the urban form not only facilitate or hinder the physical activity of city residents and their connection with nature in parks and reserves, but also reflect levels of social order in a neighbourhood. Neighbourhood disorder is a term used to describe places with high levels of perceived threat to personal safety (Ross 2011). Neighbourhood disorder reflects social disorder – the disintegration of social processes and structures that work to maintain respect and safety within a community (Kim 2008). Signs of neighbourhood disorder include dilapidated buildings, boarded-up windows, graffiti, noise, garbage, crime, public drunkenness, the public sale and/or use of illicit drugs, and the presence of delinquent youths and/or street gangs (Galea et al. 2005; Kim 2008; Ross and Mirowsky 2009). In a related concept, neighbourhood disadvantage describes places with few economic and social resources (Ross 2011); disadvantaged neighbourhoods are more likely to be disorderly.

Many studies have shown neighbourhood disorder to predict poorer mental health in residents, including increased rates of anxiety, anger and depression (reviewed by Diez Roux and Mair 2010; Ross 2011). For example, a study in New York found that people living in disorderly neighbourhoods characterized by a poor-quality built environment were 29–58% more likely to report depression in the previous six months and 36–64% more likely to report depression at any time during their life than respondents living in more orderly neighbourhoods (Galea et al. 2005). These results were adjusted for individual income, neighbourhood income, gender, age and ethnicity, indicating an effect of neighbourhood disorder separate to that of neighbourhood disadvantage or other social factors. The negative impacts of neighbourhood disorder on mental health may stem from subjective alienation – a sense of separation from others (Ross and Mirowsky 2009). This in turn may stem from a perception of powerlessness, feelings of mistrust, social isolation, and the breakdown of social norms governing acceptable behaviour (also known as normlessness; Ross 2011). Thus, despite the presence of many other people in the neighbourhood, residents are less likely to form the social bonds and support networks that create a sense of community and belonging.

Neighbourhood disorder may also predict poorer physical health, as a lack of safety or a perception of danger may discourage outdoor play by young people (Molnar et al. 2004) and exercise by adults (Fish et al. 2010). Of a sample of more than 2000 adults in Los Angeles, California, those who perceived their neighbourhood as unsafe had a body-mass index 2.81 kg/m^2 (95% CI: 0.11–5.52) higher than those who perceived their neighbourhoods as safe, controlling for gender, ethnicity and chronic disease (Fish et al. 2010). Similarly, a study in Belo Horizonte, Brazil found the probability of being overweight (BMI ≥ 25 kg/m^2) decreased with population density (reflecting the relationship between high population density and greater walkability of neighbourhoods), but increased with social vulnerability and the homicide rate within a census tract (adjusted for gender, age and lifestyle factors; Mendes et al. 2013). In Belo Horizonte, as in Los Angeles, a perception of danger in the neighbourhood may discourage outdoor physical activities such as walking; similar results have been reported from San Antonio, Texas and north-west England (Gomez et al. 2004; Harrison et al. 2007). Interestingly, the most overweight group in the study population of 3400 adults in Belo Horizonte was young people aged 18–24 years, with 77.5% of individuals overweight (Mendes et al. 2013).

6.3 Pollution and waste

The activities of humans in urban environments produce many types of pollution and waste, each with a range of potential impacts on the health and wellbeing of city dwellers. Here, I focus on air pollution and solid waste (garbage; Box 6.3).

Box 6.3 Garbage Mountains

The human population of cities generates vast quantities of garbage, including household waste (e.g., food scraps, paper, plastics and other packaging), scrap metal, electronic waste and hazardous waste. Global solid waste production was estimated at >3.5 million tonnes/day in 2010, and is predicted to rise to >6 million tonnes/day by 2025 and >12 million tonnes/day by 2100 (Hoornweg and Bhada-Tata 2012; Hoornweg et al. 2013). For a given level of affluence, city dwellers generate twice as much waste as their rural counterparts (Hoornweg et al. 2013). Centralized dumping of waste from populous cities can lead to the accumulation of enormous landfill sites or garbage mountains (Figure 6.7). Notable examples include the Payatas waste dump near Manila, the Jardim Gramacho landfill in Rio de Janeiro and the Bordo Poniente in Mexico City. Garbage mountains provide a livelihood for a section of society known as garbage pickers; adults and children who earn money by sorting through the garbage and recycling valuable items they find.

Figure 6.7 Scavengers at work on a garbage mountain in Brazil, collecting and sorting rubbish for recycling. Photograph by Marcello Casal Jr./Agência Brasil. https://commons.wikimedia.org/wiki/File%3ALixaoCatadores20080220MarcelloCasalJrAgenciaBrasil.jpg. Used under CC-BY-2.5 br http://creativecommons.org/licenses/by/2.5/br/deed.en

The garbage pickers perform an important service by retrieving and recycling a variety of plastics, metals, computer parts, printer cartridges and other electronic waste, and can earn substantially more money doing so than they would in a range of other occupations, including farm and factory work (Medina 2005, 2008; Power 2006; Samson 2008). Modern-day garbage pickers perform a similar function to the bone-pickers, bunters, toshers, sewer-hunters and other scavengers in 19th century London, who were estimated to number more than 100,000 in the 1850s (Johnson 2006). In some instances, shanty towns housing the garbage pickers spring up around a garbage mountain; these towns themselves harbour a variety of pollutants and diseases, including tetanus and tuberculosis. In 2000, a section of the Payatas dump near Manila collapsed following weeks of heavy rain, killing more than 200 people under an avalanche of waste and mud (Power 2006). However, the dump was reopened within a few months of this disaster and continues to provide a livelihood – and a home – for thousands of garbage pickers and their families.

6.3.1 Outdoor air pollution

Air pollution in cities contains a diverse mixture of gases (e.g., nitrous oxides, sulfur dioxide, carbon monoxide, ozone) and particulate matter, produced in large part by the combustion of fossil fuels for transport, industry, household heating and cooking (see Chapter 5). Particulate matter is itself a heterogeneous mixture of compounds that vary in chemical composition, size and surface area (Brook 2008). Particulate matter is generally classified by its aerodynamic diameter in μm, with the smallest particles categorized as fine (<2.5 μm; known as $PM_{2.5}$) and ultra-fine (<0.1 μm; $PM_{0.1}$). Coarse particulate matter falls in the diameter range 2.5–10 μm (PM_{10}). PM_{10} includes dust, pollen, fungal spores, soil and metal, and has an average lifespan of hours to days and a distribution of 10–100 km from the source (Brook 2008). $PM_{2.5}$ is composed of elemental carbon, hydrocarbons, organic compounds and metals, has an average lifespan of days to weeks, and is distributed regionally (≥1000 km from the source). $PM_{0.1}$ has a similarly mixed chemical composition but an average lifespan of just minutes to hours and a distribution of a few hundred metres from the source (Brook 2008). Thus, the distribution and composition of particulate pollution can vary widely across a city, with the highest concentrations of $PM_{0.1}$ and $PM_{2.5}$ found close to busy roads.

Both short- and long-term exposures to particulate matter in urban environments is associated with an increased risk of cardiovascular disease in humans, including cardiac ischaemia, heart attacks, acute heart failure, stroke and peripheral arterial disease (Brook 2008). For example, the risk of heart attack may be increased by a factor of approximately 2.7 after only one hour of exposure to traffic (Peters et al. 2004). An increase in long-term $PM_{2.5}$ exposure of 10 $\mu g/m^3$ was associated with a 76% increase in risk of mortality from cardiovascular disease in a study of >65,000 healthy, post-menopausal women in the USA (Miller et al. 2007). Exposure to particulate pollution also increases the incidence of respiratory illnesses, including asthma, bronchitis and lung cancer, even among non-smokers (Valavanidis et al. 2008). Available evidence indicates a greater role of fine and ultra-fine particles ($PM_{2.5}$ and $PM_{0.1}$) in cardiovascular and other diseases because these particles are small enough to penetrate deeply into the airways of the respiratory tract, where they may lodge in the parenchyma (the parts of the lung involved in gas exchange; Valavanidis et al. 2008). Adverse health effects of fine-particulate air pollution depend on both the concentration of particles in the atmosphere and length of exposure, with long-term exposure having greater and more persistent effects (Pope 2007). Approximately 3.2 million (uncertainty interval 2.8–3.6 million) deaths and 76 million disability-adjusted life-years were attributed to ambient particulate-matter pollution worldwide in 2010, with this risk factor ranked ninth globally for its attributable burden of disease (Lim et al. 2012). Disability-adjusted life-years represent the sum of years lived with a disability due to illness plus the years of life lost due to premature death (Jordan et al. 2014).

6.3.2 Indoor air pollution

Despite the more obvious concern surrounding outdoor air pollution in cities, indoor air pollution is actually responsible for a greater burden of disease worldwide. Household air pollution from the combustion of solid fuels was responsible for an estimated 3.5 million deaths and 108 million disability-adjusted life-years worldwide in 2010; this risk factor was ranked third globally for its attributable burden of disease, behind only high blood pressure and tobacco smoking (including secondhand smoke; Lim et al. 2012). On a regional basis, household air pollution causes the greatest health burden in Asia and sub-Saharan Africa, where biomass fuels such as wood, charcoal, dung and crop residues are commonly used for cooking and many dwellings are poorly ventilated (Saksena et al. 2003; Bailis et al. 2005; Dasgupta et al. 2006; Zhou et al., 2011). Without a substantial change in practice (e.g., a transition to cleaner-burning fuels such as charcoal or LPG), the indoor air pollution resulting from the use of biomass fuels is predicted to cause 9.8 million premature deaths in Africa between 2000 and 2030 (Bailis et al. 2005). This estimate includes 8.1 million deaths among children <5 years of age from lower respiratory infections and 1.7 million deaths among adult women from chronic obstructive pulmonary disease.

In developed countries, indoor air pollution caused by hazardous emissions from household and office products, such as formaldehyde and volatile organic compounds (VOCs), has been gaining greater research attention (Zhang and Smith 2003; Steinemann 2009). Volatile organic compounds are organic chemicals with a low boiling point and thus a high vapour pressure at room temperature, such that large numbers of molecules evaporate (from liquid sources) or sublime (from solid sources) into the air. These compounds are commonly found in paint, glue, plastics, synthetic fibres and solvents, and may reach higher concentrations indoors than outdoors (Mølhave 1986; Dales et al. 2008). Certain VOCs such as benzene and 1,3butadiene are known to be carcinogenic in humans, and indoor exposure to VOCs is associated with an increased relative risk of leukaemia and lymphoma (Viegi et al. 2004; Irigarary et al. 2007). Fragranced household products- such as air fresheners, dishwashing and laundry detergents and fabric softeners- also contain a wide range of VOCs, some of which are classified as toxic or hazardous (e.g., acetaldehyde and chloromethane; Steinemann 2009). Use of fragranced products is associated with a variety of health problems including headaches, asthma and contact allergies, but the majority of the VOCs they contain and emit are not listed on product ingredients, making identification of problematic ingredients difficult (Steinemann et al. 2011).

6.4 Climatic changes in urban environments

The urban heat-island effect can exacerbate high diurnal and nocturnal temperatures during summer heatwaves, making conditions more unpleasant and more dangerous for city residents. Hard city surfaces heat up quickly during the day and then release their heat over an extended period at night, subjecting urban residents to sustained thermal stress during periods of hot weather, and increasing rates of morbidity and mortality (Weng and Yang 2003; Harlan et al. 2006; Kalkstein et al. 2013; Azhar et al. 2014; Hondula and Barnett 2014). Many different causes of death increase in frequency during heatwaves, including cardiac arrest, stroke and respiratory failure (Kovats and Hajat 2008; Giua et al. 2010; Tan et al, 2010; Chen et al. 2013; Kalkstein et al. 2013).

For example, the urban heat-island effect in Shanghai, China contributed to substantial heat-related mortality during a 9-day heatwave in August 1998 (Tan et al. 2010). The all-cause, excess mortality rate in the centre of Shanghai was 27.3/100,000 people, compared with 7/100,000 in exurban (rural) districts. The number of excess deaths per day increased as the heatwave progressed, peaking at 453 on the ninth and final day (Tan et al. 2010). A study in Brisbane, Australia, found a 7.2% (95% CI: 4.7–9.8%) increase in hospital admissions on the day following a 10 °C increase in daily maximum temperature during the summer (Hondula and Barnett 2014), while another observed a 15% (5–26%) increase in the rate of hospital admissions for haemorrhagic stroke for each 1°C increase in daily maximum temperature (Wang et al. 2009). Certain sub-groups of the urban population, including the elderly, the socially disadvantaged and the homeless, are at greatest risk of illness and death due to elevated urban temperatures (Harlan et al. 2006; Kovats and Ebi 2006; Kovats and Hajat 2008; Loughnan et al. 2012).

Elevated temperatures also influence human behaviour in cities; field studies and laboratory experiments have shown that high temperatures increase irritability and aggression (sometimes known as the heat effect; reviewed by Anderson 2001). Violent crime increases with temperature (Anderson 2001; Anderson and Delisi 2011). Based on data from 50 cities in the USA between 1950 and 2008, the incidence of violent crime (murder and serious assault) was higher in hotter years, increasing by an average of 142 crimes/100,000 people for every 1°C increase in annual mean temperature (equivalent to 79 crimes/100,000 people for every 1°F increase; Anderson and DeLisi 2011). Crime also varies with temperature from month to month. In St Louis, Missouri, violent crimes increased by an average of 9.9 incidents (or 1.3%) per month for each 1°C increase in monthly temperature above expected (temperature anomaly; Mares 2013). Normal temperature variation in St. Louis between the coldest and warmest months (January and August) is close to 28 °C; this translates to a predicted seasonal fluctuation in violent crime of ~35%.

Extreme heat and its adverse effects on urban populations are expected to increase under climate change, including the number and length of heatwaves (IPCC 2014) and rates of aggression and crime (Anderson and DeLisi 2011; Mares 2013). Combined with the urban heat-island effect, more frequent heatwaves will lead to increased heat-related illness and mortality in many cities around the world (Huang et al. 2011a; IPCC 2014). In addition, a 2 °C (3.6°F) increase in mean annual temperature (a feasible scenario for much of North America by 2025; IPCC 2014) could translate into 50,000 more violent crimes (murders and serious assaults) per year in the USA (Anderson and DeLisi 2011). In addition to the substantial personal and societal costs, the economic cost of this additional burden of disease, violence and death will be high. The cost to society of violent crimes has been estimated at US$1.28 million per murder and US$19,537 per aggravated assault (McCollister et al. 2010). Thus, if 10% of the additional violent crimes per year in the USA attributable to a 2 °C increase in mean annual temperature were murders, this would translate into an additional cost to society of $7.28 billion.

6.5 Health inequities in the world's cities

While people who live in urban areas tend to have better health and greater life expectancy than their rural counterparts, these broad trends mask grave inequities in human health across the urban socioeconomic spectrum. For example, poor urban women are less likely to have a skilled birth attendant present during labour, leading to a higher risk of complications that can result in maternal disability or death (WHO & UN-Habitat 2010). In Bangladesh, rates of skilled birth attendance range from 6% for the poorest urban women (bottom quintile) to 75% for the richest (top quintile). A study across 21 countries revealed that urban women are 50% more likely than urban men and 80% more likely than rural women to be HIV positive, with the problem compounded by low socioeconomic status (WHO & UN-Habitat 2010). Based on recent data from Demographic and Health Surveys (2003–2008), more than 35% of urban women in Swaziland are HIV positive (WHO & UN-Habitat 2010).

Household wealth also plays a major role in child malnutrition and infant and child mortality in cities around the world. Globally, chronic malnutrition causes more than a third of all deaths during childhood; while malnutrition among children under 5 years is less common in urban than in rural areas, there is substantial variation within the urban population as a whole. Across Africa, Asia and the Americas, the poorest urban children are three times more likely than the richest to have their growth stunted by insufficient nutrition; the risk of chronic child malnutrition decreases progressively as family income rises (WHO & UN-Habitat 2010). Within each of these three regions, children from the poorest urban families are approximately twice as likely to die before the age of 5 as children from the richest urban families. Africa has the highest rates of child mortality (>130/1000 live births for the poorest urban families; data from 2000–2007).

The health burden of the urban poor – disproportionately borne by women and children – has a range of causes, including poor-quality housing, inadequate sanitation and insufficient access to clean drinking water; high population densities in slums and other informal settlements that assist the transmission of infectious diseases spread by person to person contact, such as lower-respiratory infections and tuberculosis; poor access to health care, higher rates of violent crime and a higher prevalence of risk-taking behaviours (Zulu et al. 2002; Campbell and Campbell 2007; COHRE 2008; Montgomery 2009; WHO & UN-Habitat 2010; Mendes et al. 2013). The insecure nature of dwellings and shared toilet facilities in informal settlements exposes women to an increased risk of sexual assault and sexually transmitted infections including HIV (Kalichman and Simbayi 2004; Amuyunzu-Nyamongo et al. 2007; COHRE 2008; Amnesty International 2010). Poverty, homelessness and food insufficiency may also lead to women exchanging sex for food or accommodation, which in turn increases their risk of contracting sexually transmitted infections (Weiser et al. 2007).

Residents of slums and other informal settlements suffer from substantial uncertainty of tenure and the possibility of forced evictions from their homes through "slum clearance" (Greene 2003; Padhi 2007; COHRE 2008; Amnesty International 2010; Shin and Li 2012; Steinbrink 2013). A notable recent example of forced evictions from informal settlements was Operation Murambatsvina (also known as Operation Restore Order or Operation Drive Out Rubbish), launched by President Mugabe of Zimbabwe in 2005. The operation involved the rapid clearance of slums by the police and military, first in Harare and then a range of other cities throughout the country. An estimated 700,000 people lost their homes, their source of livelihood or both during the operation, which drove hundreds of thousands of the urban poor deeper into poverty and destitution (Tibaijuka 2005). Slum clearance is often associated with large construction projects for major world events such as the Olympic Games and the Football World Cup, along with a desire to "clean up" the image of the city that is hosting the event (Greene 2003; Power 2006; Shin and Li 2012). The urban poor also tend to suffer most from natural disasters, such as cyclones and floods, with this trend set to be exacerbated under climate change (Anderson and DeLisi 2011; IPCC 2014).

6.6 Summary

The processes of urbanization have wide-ranging impacts on the humans who live in cities; from individuals to families, communities and the entire population. While cities can offer much to their residents – opportunity, employment, culture, energy-efficient living and a sense of community – they can also be places of grime, crime, illness and despair. Characteristics of the urban form can promote or hinder health and wellbeing in human populations. Neighbourhood design and a sense of order or disorder influence exercise patterns and levels of overweight, obesity, social connection/isolation and depression. Access to neighbourhood parks improves both mental and physical health while facilitating a continuing connection between city residents and nature; urban sprawl encourages car-dependence and can foster a sense of isolation from community. Cities produce vast quantities of pollution, waste and heat, and may also facilitate the persistence and spread of infectious diseases. In cities around the world, it is the urban poor who live most closely with these unpleasant products of urbanization; contrasts between the health of advantaged and disadvantaged groups in urban society are stark. However, cities are the way of the present and the future – there is no going back to an agrarian lifestyle. We must adapt our cities to counteract the problems of pollution, waste, heat, sprawl, separation from nature and isolation from community for all urban residents, not just the wealthy and the well connected. In the next chapter, I discuss practical strategies for creating healthier, more liveable urban spaces – for people and for all the other species with which we share our cities.

Study questions

1. There can be few opportunities for humans to experience nature in cities. How would you increase these opportunities in a city or town that you know?
2. Access to parks and other green open spaces in cities is important for human health – discuss.
3. Using examples, describe the range of human–wildlife interactions that occur in urban environments.
4. Define the following terms: walkability, urban sprawl and neighbourhood disorder. How do they influence human health and wellbeing in cities?
5. How do climatic changes associated with urbanization affect urban-dwelling humans?
6. Which is a greater threat to human health in cities – indoor or outdoor air pollution? Why?
7. Describe the main factors that contribute to health inequities in the world's cities. Who bears the greatest burden of this inequity?

References

Adams, A.L., Dickinson, K.J.M., Robertson, B.C. *et al.* (2013) Predicting summer site occupancy for an invasive species, the common brushtail possum (*Trichosurus vulpecula*), in an urban environment. *PLoS ONE* **8**, e58422.

Allen, A.P. and O'Conner, R.J. (2000) Hierarchical correlates of bird assemblage structure on northeastern U.S.A. lakes. *Environmental Monitoring and Assessment*, **62**, 15–37.

Amnesty International (2001) *Insecurity and Indignity: Women's Experiences in the Slums of Nairobi, Kenya*, Amnesty International Publications, London.

Amuyunzu-Nyamongo, M., Okeng'o, L., Wagura, A. *et al.* (2007) Putting on a brave face: the experience of women living with HIV and AIDS in informal settlements of Nairobi, Kenya. *AIDS Care*, **19**, 25–34.

Anderson, C.A. (2001) Heat and violence. *Current Directions in Psychological Science*, **10**, 33–38.

Anderson, C.A. and DeLisi, M. (2011) Implications of global climate change for violence in developed and developing countries, in *The Psychology of Social Conflict and Aggression* (eds J. Forgas, A. Kruglanski, and K. Williams), Psychology Press, New York, pp. 249–265.

Azhar, G.S., Mavalankar, D., Nori-Sarma, A. *et al.* (2014) Heat-Related mortality in India: excess all-cause mortality associated with the 2010 Ahmedabad Heat Wave. *PLoS ONE*, **9** (e), 91831.

Bailis, R., Ezzati, M., and Kammen, D.M. (2005) Mortality and greenhouse gas impacts of biomass and petroleum energy futures in Africa. *Science*, **308**, 98–103.

Banks, W.A., Altmann, J., Sapolsky, R.M. *et al.* (2003) Serum leptin levels as a marker for a syndrome X-like condition in wild baboons. *The Journal of Clinical Endocrinology and Metabolism*, **88**, 1234–1240.

Barton, J. and Pretty, J. (2010) What is the best dose of nature and green exercise for improving mental health? A multi-study analysis. *Environmental Science and Technology*, **44**, 3947–3955.

Bateman, P.W. and Fleming, P.A. (2012) Big city life: carnivores in urban environments. *Journal of Zoology*, **287**, 1–23.

Bates, D.G. (2012) On forty years: remarks from the editor. *Human Ecology*, **40**, 1–4.

Bauman, A.E., Reis, R.S., Sallis, J.F. *et al.* (2012) Correlates of physical activity: why are some people physically active and others not? *The Lancet*, **380**, 258–271.

Beckmann, J.P. and Berger, J. (2003) Rapid ecological and behavioural changes in carnivores: the response of black bears (*Ursus americanus*) to altered food. *The Zoological Society of London*, **261**, 207–212.

Beier, P. (1991) Cougar attacks on humans in the United States and Canada. *Wildlife Society Bulletin*, **19**, 403–412.

Bhatia, S., Athreya, V., Grenyer, R. *et al.* (2013) Understanding the role of representations of human-leopard conflict in Mumbai through media-content analysis. *Conservation Biology*, **27**, 588–594.

Brook, R.D. (2008) Cardiovascular effects of air pollution. *Clinical Science*, **115**, 175–187.

Campbell, M. (2008) An animal geography of avian feeding habits in Peterborough, Ontario. *Area*, **40**, 472–480.

Campbell, T. and Campbell, A. (2007) Emerging disease burdens and the poor in cites of the developing world. *Journal of Urban Health: Bulletin of the New York Academy of Medicine*, **84**, 54–64.

Carrus, G., Scopelliti, M., Lafortezza, R. *et al.* (2015) Go green, feel better? The positive effects of biodiversity on the well-being of individuals visiting urban and peri-urban green areas. *Landscape and Urban Planning*, **134**, 221–228.

Casagrande, S.S., Gittelsohn, J., Zonderman, A.B. *et al.* (2011) Association of walkability with obesity in Baltimore City, Maryland. *American Journal of Public Health*, **101**, 318–324.

Chamberlain, D.E., Vickery, J.A., Glue, D.E. *et al.* (2005) Annual and seasonal trends in the use of garden feeders by birds in winter. *Ibis*, **147**, 563–575.

Chapman, R. and Jones, D.N. (2009) Just feeding the ducks: quantifying a common wildlife-human interaction. *The Sunbird*, **39**, 19–28.

Chapman, R. and Jones, D.N. (2011) Foraging by native and domestic ducks in urban lakes: behavioural implications of all that bread. *Corella*, **35**, 101–106.

Cervero, R. and Kockelman, K. (1997) Travel demand and the 3Ds: density, diversity, and design. *Transportation Research Part D - Transport and Environment*, **2**, 199–219.

Cervero, R., Sarmiento, O.L., Jacoby, E. *et al.* (2009) Influences of built environments on walking and cycling: lessons from Bogota. *International Journal of Sustainable Transportation*, **3**, 203–226.

Chaudhry, P., Bagra, K., and Singh, B. (2011) Urban greenery status of some Indian cities: a short communication. *International Journal of Environmental Science and Development*, **2**, 98–101.

Chen, R., Wang, C., Meng, X. *et al.* (2013) Both low and high temperature may increase the risk of stroke mortality. *Neurology*, **81**, 1064–1070.

Chiesura, A. (2004) The role of urban parks for the sustainable city. *Landscape and Urban Planning*, **68**, 129–138.

Choudhury, A. (2004) Human-elephant conflicts in Northeast India. *Human Dimensions of Wildlife*, **9**, 261–270.

Cilliers, J. (2008) *Africa in the New World: How Global and Domestic Developments will Impact by 2025*. Institute for Security Studies Monographs, no. 151. Institute for Security Studies, Pretoria.

Clark, N.E., Lovell, R., Wheeler, B.W. *et al.* (2014) Biodiversity, culture pathways, and human health: a framework. *Trends in Ecology and Evolution*, **29**, 198–204.

Clarke, L.W., Jenerette, G.D., and Davalia, A. (2013) The luxury of vegetation and the legacy of tree biodiversity in Los Angeles, CA. *Landscape and Urban Planning*, **116**, 48–59.

COHRE (2008) *Women, Slums and Urbanization: Examining the Causes and Consequences*, Centre on Housing Rights and Evictions, Geneva.

Coley, R.L., Kuo, F.E., and Sullivan, W.C. (1997) Where does community grow? The social context created by nature in urban public housing. *Environment and Behavior*, **29**, 468–494.

Dales, R., Liu, L., Wheeler, A.J. *et al.* (2008) Quality of indoor residential air and health. *Canadian Medical Association Journal*, **179**, 147–152.

Daniels, G.D. and Kirkpatrick, J.B. (2006) Does variation in garden characteristics influence the conservation of birds in suburbia? *Biological Conservation*, **133**, 326–335.

Dasgupta, S., Huq, M., Khaliquzzaman, M. *et al.* (2006) Indoor air quality for poor families: new evidence from Bangladesh. *Indoor Air*, **16**, 426–444.

Dieleman, F. and Wegener, M. (2004) Compact city and urban sprawl. *Built Environment*, **30**, 308–323.

Diez Roux, A.V. and Mair, C. (2010) Neighborhoods and health. *Annals of the New York Academy of Sciences*, **1186**, 125–145.

Escobedo, F.J., Nowak, D.J., Wagner, J.E. *et al.* (2006) The socioeconomics and management of Santiago de Chile's public urban forests. *Urban Forestry and Urban Greening*, **4**, 105–114.

Ewing, R., Pendall, R., and Chen, D. (2003) Measuring sprawl and its transportation impacts. *Transportation Research Record: Journal of the Transportation Research Board*, **1**, 175–183.

Farr, D. (2012) *Sustainable Urbanism: Urban Design with Nature*, John Wiley & Sons Inc., Hoboken.

Fish, J.S., Ettner, S., Ang, A. *et al.* (2010) Association of perceived neighbourhood safety on body mass index. *American Journal of Public Health*, **100**, 2296–2303.

Frank, L.D., Anderson, M.A., and Schmid, T.L. (2004) Obesity relationships with community design, physical activity, and time spent in cars. *American Journal of Preventive Medicine*, **27**, 87–96.

Fuller, R.A., Irvine, K.N., Devine-Wright, P. *et al.* (2007) Psychological benefits of greenspace increase with biodiversity. *Biology Letters*, **3**, 390–394.

Fuller, R.A., Warren, P.H., Armsworth, P.R. *et al.* (2008) Garden bird feeding predicts the structure of urban avian assemblages. *Diversity and Distributions*, **14**, 131–137.

Galea, S., Ahern, J., Rudenstine, S. *et al.* (2005) Urban built environment and depression: a multilevel analysis. *Journal of Epidemiology and Community Health*, **59**, 822–827.

Garden, F.L. and Jalaludin, B.B. (2009) Impact of urban sprawl on overweight, obesity, and physical activity in Sydney, Australia. *Journal of Urban Health*, **86**, 19–30.

Germano, J.M. and Bishop, P.J. (2009) Suitability of amphibians and reptiles for translocation. *Conservation Biology*, **23**, 7–15.

Giles-Corti, B., Broomhall, M.H., Knuiman, M. *et al.* (2005) Increasing walking. How important is distance to, attractiveness, and size of public open space? *American Journal of Preventive Medicine*, **28**, 169–176.

Giles-Corti, B., Macaulay, G., Middleton, N. *et al.* (2015) Developing a research and practice tool to measure walkability: a demonstration project. *Health Promotion Journal of Australia*, **25**, 160–166.

Giua, A., Abbas, M.A., Murgia, N. *et al.* (2010) Climate and stroke: a controversial association. *International Journal of Biometeorology*, **54**, 1–3.

Glue, D. (2006) Variety at winter bird tables. *Bird Populations*, **7**, 212–215.

Gomez, J.E., Johnson, B.A., Selva, M., and Sallis, J.F. (2004) Violent crime and outdoor activity among inner-city youth. *Preventive Medicine*, **39**, 876–881.

Grahn, P. and Stigsdotter, U.A. (2003) Landscape planning and stress. *Urban Forestry and Urban Greening*, **2**, 1–18.

Green, M. (2014) Little Fox, Big Problem. *The Age*, July 4th, 2014. http://www.theage.com.au/action/printArticle?id=5567201.

Greene, S.J. (2003) Staged cities: mega-events, slums clearance, and global capital. *Yale Human Rights and Development Law Journal*, **6**, 161–187.

Hallal, P.C., Andersen, L.B., Bull, F. *et al.* (2012) Global physical activity levels: surveillance progress, pitfalls, and prospects. *The Lancet*, **380**, 247–257.

Hammer, M.S., Swinburn, T.K., and Neitzel, R.L. (2014) Environmental noise pollution in the United States: developing an effective public health response. *Environmental Health Perspectives*, **122**, 115–119.

Hankey, S., Marshall, J.D., and Brauer, M. (2012) Health impacts of the built environment: within-urban variability in physical inactivity, air pollution, and ischemic heart disease mortality. *Environmental Health Perspectives*, **120**, 247–253.

Haq, S.M.A. (2011) Urban green space and an integrative approach to sustainable environment. *Journal of Environmental Protection*, **2**, 601–608.

Harlan, S.L., Brazel, A.J., Prashad, L. *et al.* (2006) Neighborhood microclimates and vulnerability to heat stress. *Social Science and Medicine*, **63**, 2847–2863.

Harper, M.J., McCarthy, M.A., and van der Ree, R. (2008) Resources at the landscape scale influence possum abundance. *Austral Ecology*, **33**, 243–252.

Harrison, R.A., Gemell, I., and Heller, R.F. (2007) The population effect of crime and neighbourhood on physical activity: an analysis of 15,461 adults. *Journal of Epidemiology and Community Health*, **61**, 34–39.

Hes, D. and Plessis, C.D. (2015) *Designing for Hope: Pathways to Regenerative Sustainability*, Routledge, Abingdon.

Heynen, N.C. (2003) The scalar production of injustice within the urban forest. *Antipode*, **35**, 980–998.

Hondula, D.M. and Barnett, A.G. (2014) Heat-related morbidity in Brisbane, Australia: spatial variation and area-level predictors. *Environmental Health Perspectives*, **122**, 831–836.

Hoornweg, D., and Bhada-Tata, P. (2012) *What a Waste: A Global Review of Solid Waste Management*. World Bank, Washington DC. http://siteresources.worldbank.org/INTURBANDEVELOPMENT/Resources/336387-1334852610766/What_a_Waste2012_Final.pdf (accessed 10/06/2015)

Hoornweg, D., Bhada-Tata, P., and Kennedy, C. (2013) Waste production must peak this century. *Nature*, **502**, 615–617.

Huang, C., Barnett, A.G., Wang, X. *et al.* (2011a) Projecting future heat-related mortality under climate change scenarios: a systematic review. *Environmental Health Perspectives*, **119**, 1681–1690.

Huang, G., Zhou, W., and Cadenasso, M.L. (2011b) Is everyone hot in the city? Spatial pattern of land surface temperatures, land cover, and neighborhood socioeconomic characteristics in Baltimore, MD. *Journal of Environmental Management*, **92**, 1753–1759.

IPCC. 2014. *Climate Change 2014: Synthesis Report. Contribution of Working Groups I, II and III to the Fifth Assessment Report of the Intergovernmental Panel on Climate Change*. Pachauri, R.K., and L.A. Meyer (eds). IPCC, Geneva.

Irigarary, P., Newby, J.A., Clapp, R. *et al.* (2007) Lifestyle-related factors and environmental agents causing cancer: an overview. *Biomedicine and Pharmacotherapy* **61**, 640-658.

Johnson, S. (2006) *The Ghost Map: The Story of London's Most Terrifying Epidemic, and How it Changed Science, Cities, and the Modern World*, Riverhead, New York.

Jones, D.N. and Reynolds, S.J. (2008) Feeding birds in our towns and cities: a global research opportunity. *Journal of Avian Biology*, **39**, 265–271.

Jordan, H., Dunt, D., Hollingsworth, B. *et al.* (2014) Costing the morbidity and mortality consequences of zoonoses using health-adjusted life years. *Transboundary and Emerging Diseases*. doi: 10.1111/tbed.12305

Kaczynski, A.T. and Henderson, K.A. (2007) Environmental correlates of physical activity: a review of evidence about parks and recreation. *Leisure Science*, **29**, 315–354.

Kalichman, S.C. and Simbayi, L.C. (2004) Sexual assault history and risk for sexually transmitted infections among women in an African township in Cape Town, South Africa. *AIDS Care*, **16**, 681–689.

Kalkstein, L.S., Sailor, D., Shickman, K. *et al.* (2013) *Assessing the health impacts of urban heat island reduction strategies in the District of Columbia*. Global Cool Cities Alliance, Washington DC. http://www.coolrooftoolkit.org/wp-content/uploads/2013/12/DC-Heat-Mortality-Study-for-DDOE-FINAL.pdf (accessed on 12/06/2015)

Kennedy, C., Cuddihy, J., and Engel-Yan, J. (2007) The changing metabolism of cities. *Journal of Industrial Ecology*, **11**, 43–59.

Kennedy, C., Pincetl, S., and Bunje, P. (2011) The study of urban metabolism and its applications to urban planning and design. *Environmental Pollution*, **159**, 1965–1973.

Kim, D. (2008) Blues from the neighborhood? Neighborhood characteristics and depression. *Epidemiologic Reviews*, **30**, 101–107.

Kingsley, J.Y. and Townsend, M. (2006) 'Dig in' to social capital: community gardens as mechanisms for growing urban social connectedness. *Urban Policy and Research*, **24**, 525–537.

Knopfler, M. (1982) *Telegraph Road*, Vertigo/Universal Music UK, London.

Kovats, R.S. and Ebi, K.L. (2006) Heatwaves and public health in Europe. *European Journal of Public Health*, **16**, 592–599.

Kovats, R.S. and Hajat, S. (2008) Heat stress and public health: a critical review. *The Annual Review of Public Health*, **29**, 1–9.

Kuo, F.E. (2001) Coping with poverty: impacts of environment and attention in the inner city. *Environment and Behavior*, **33**, 5–34.

Lamarque, F., Anderson, J., Fergusson, R. *et al.* (2008) *Human-Wildlife Conflict in Africa: Causes, Consequences, and Management Strategies*. FAO Forestry Paper 157. Food and Agriculture Organization of the United Nations, Rome. http://www.fao.org/docrep/012/i1048e/i1048e00.htm (accessed 10/06/2015)

Lee, K.E., Williams, K.J.H., Sargent, L.D. *et al.* (2015) 40-second green roof views sustain attention: the role of micro-breaks in attention restoration. *Journal of Environmental Psychology*, **42**, 182–189.

Lim, S.S., Vos, T., Flaxman, A.D. *et al.* (2012) A comparative risk assessment of burden of disease and injury attributable to 67 risk factors and risk factor clusters in 21 regions, 1990-2010: a systematic analysis for the Global Burden of Disease Study 2010. *Lancet*, **380**, 2224–2260.

Linnell, J.D., Aanes, R., and Swenson, J.E. (1997) Translocation of carnivores as a method for managing problem animals: a review. *Biodiversity and Conservation*, **6**, 1245–1257.

Lopez, R. (2004) Urban sprawl and risk for being overweight or obese. *American Journal of Public Health*, **94**, 1574–1579.

Loughnan, M., Nicholls, N., and Tapper, N.J. (2012) Mapping heat health risks in urban areas. *International Journal of Population Research*, **2010**, 1–12.

Lowe, M., Boulange, C., and Giles-Corti, B. (2014) Urban design and health: progress to date and future challenges. *Health Promotion Journal of Australia*, **25**, 14–18.

Maas, J., van Dillen, S.M.E., Verheij, R.A. *et al.* (2009) Social contact as a possible mechanism behind the relationship between green space and health. *Health and Place*, **15**, 586–595.

Maller, C., Townsend, M., Pryor, A. *et al.* (2004) Healthy nature healthy people: 'Contact with nature' as an upstream health promotion intervention for populations. *Health Promotions International*, **21**, 45–54.

Mares, D. (2013) Climate change and crime: monthly temperature and precipitation anomalies and crime rates in St. Louis, MO 1990-2009. *Crime, Law, and Social Change*, **59**, 185–208.

Martinez-Fernandez, C., Audirac, I., Fol, S., and Cunningham-Sabot, E. (2012) Shrinking cities: urban challenges of globalization. *International Journal of Urban and Regional Research*, **36**, 213–225.

Mason, P.R. (2009) Zimbabwe experiences the worst epidemic of cholera in Africa. *The Journal of Infection in Developing Countries*, **3**, 148–151.

McCollister, K.E., French, M.T., and Fang, H. (2010) The cost of crime to society: new crime-specific estimates for policy and program evaluation. *Drug and Alcohol Dependence*, **108**, 98–109.

McKee, D. (2003) Cougar attacks on humans: a case report. *Wilderness and Environmental Medicine*, **14**, 169–173.

Medina, M. (2005) Serving the unserved: Informal refuse collection in Mexican cities. *Waste Management and Research*, **23**, 390–397.

Medina, M. (2008) The informal recycling sector in developing countries: organizing waste pickers to enhance their impact. *Gridlines Note* no. 44. Public-Private Infrastructure Advisory Facility (PPIAF), World Bank, Washington DC. https://www.ppiaf.org/sites/ppiaf.org/

files/publication/Gridlines-44-Informal%20Recycling%20-%20MMedina.pdf (accessed 12/06/2015)

Mendes, L.L., Nogueira, H., Padez, C. *et al.* (2013) Individual and environmental factors associated for overweight in urban population of Brazil. *BMC Public Health*, **13**, 988. doi: 10.1186/1471-2458-13-988

Miller, J.R. (2005) Biodiversity conservation and the extinction of experience. *Trends in Ecology and Evolution*, **20**, 430–343.

Miller, K.A., Siscovick, D.S., Sheppard, L. *et al.* (2007) Long-term exposure to air pollution and incidence of cardiovascular events in women. *The New England Journal of Medicine*, **356**, 447–458.

Mølhave, L. (1986) Indoor air quality in relation to sensory irritation due to volatile organic compounds. *ASHRAE Transactions*, **92**, 306–316.

Molnar, B.E., Gortmaker, S.L., Bull, F.C. *et al.* (2004) Unsafe to play? Neighborhood disorder and lack of safety predicts reduced physical activity among urban children and adolescents. *American Journal of Health Promotion*, **18**, 378–379.

Montgomery, M. 2009. Urban poverty and health in developing countries. Population Bulletin **64**(2).

Mukandavire, Z., Liao, S., Wang, J. *et al.* (2011) Estimating the reproductive numbers for the 2008-2009 cholera outbreaks in Zimbabwe. *Proceedings of the National Academy of Sciences of the United States of America*, **108**, 8767–8772.

NSW Government. (2004) *Planning Guidelines for Walking and Cycling*. Department of Infrastructure, Planning, and Natural Resources, Sydney. http://www.planning.nsw.gov.au/plansforaction/pdf/guide_pages.pdf (accessed 15/06/2015)

Padhi, R. (2007) Forced evictions and factory closures: rethinking citizenship and rights of working class women in Delhi. *Indian Journal of Gender Studies*, **14**, 74–92.

Parsons, H., Major, R.E., and French, K. (2006) Species interactions and habitat associations of birds inhabiting urban areas of Sydney, Australia. *Austral Ecology*, **31**, 217–227.

Pedlowski, M.A., Da Silva, V.A.C., Adell, J.J.C. *et al.* (2002) Urban forest and environmental inequality in Campos dos Goytacazes, Rio de Janeiro, Brazil. *Urban Ecosystems*, **6**, 9–20.

Peters, A., von Klot, S., Heier, M. *et al.* (2004) Exposure to traffic and the onset of myocardial infarction. *The New England Journal of Medicine*, **351**, 1721–1730.

Pietsch, R.S. (1994) The fate of urban Common Brushtail Possums translocated to sclerophyll forest, in *Reintroduction Biology of Australian and New Zealand Fauna* (ed M. Serena), Surrey Beatty and Sons, Chipping Norton, pp. 239–246.

Pikora, T., Giles-Corti, B., Bull, F. *et al.* (2003) Developing a framework for assessment of the environmental determinants of walking and cycling. *Social Science and Medicine*, **56**, 1693–1703.

Pope, A.C. (2007) Mortality effects on longer term exposures to fine particulate air pollution: review of recent epidemiological evidence. *Inhalation Toxicology*, **19**, 33–38.

Power, M. (2006) The magic mountain: trickle-down economics in a Philippine garbage dump. *Harpers*, **1879**, 57–71.

Pyle, R.M. (1978) The extinction of experience. *Horticulture*, **56**, 64–67.

Redman, C.L. and Jones, N.S. (2005) The environmental, social, and health dimensions of urban expansion. *Population and Environment*, **26**, 505–520.

Redman, C.L. (1999) *Human Impacts on Ancient Environments*, University of Arizona Press, Tucson.

Ross, C.E. (2011) Collective threat, trust, and the sense of personal control. *Journal of Health and Social Behaviour*, **52**, 287–296.

Ross, C.E. and Mirowsky, J. (2009) Neighborhood disorder, subjective alienation, and distress. *Journal of Health and Social Behavior*, **50**, 49–64.

Saldivar-Tanaka, L. and Krasny, M.E. (2004) Culturing community development, neighborhood open space, and civic agriculture: the case of Latino community gardens in New York City. *Agriculture and Human Values*, **21**, 399–412.

Saelens, B.E., Sallis, J.F., and Frank, L.D. (2003) Environmental correlates of walking and cycling: findings from the transportation, urban design, and planning literatures. *Environment and Physical Activity*, **25**, 80–91.

Saksena, S., Singh, P.B., Kumar Prasad, R. *et al.* (2003) Exposure of infants to outdoor and indoor air pollution in low-income urban areas – a case study of Delhi. *Journal of Exposure Analysis and Environmental Epidemiology*, **13**, 219–230.

Šálek, M., Drahnikova, L., and Tkadlec, E. (2015) Changes in home-range sizes and population densities of carnivore species along the natural to urban habitat gradient. *Mammal Review*, **45**, 1–14.

Sallis, J., Millstein, R., and Carlson, J. (2011) Community design for physical activity, in *Making Healthy Places: Designing and Building for Health, Well-Being, and Sustainability* (eds A. Dannenberg, H. Frumkin, and R. Jackson), Island Press, Washington, DC, pp. 33–49.

Sallis, J.F., Saelens, B.E., Frank, L.D. *et al.* (2009) Neighborhood built environment and income: examining multiple health outcomes. *Social Science & Medicine*, **68**, 1285–1293.

Samson, M. (2008) *Refusing to be Cast Aside: Waste Pickers Organising around the World*, Women in Informal Employment: Globalizing and Organizing (WIEGO), Cambridge, Massachusetts.

Saunders, G., Coman, B., Kinnear, J. *et al.* (1995) *Managing Vertebrate Pests: Foxes*. Bureau of Resource Sciences, Australian Government, Canberra. http://www.southwestnrm.org.au/sites/default/files/uploads/ihub/saunders-g-et-al-1995-managing-vertebrate-pests-foxes-table-contents.pdf (accessed (15/04/2014)

Schell, L.M. and Denham, M. (2003) Environmental pollution in urban environments and human biology. *Annual Review of Anthropology*, **32**, 111–134.

Silver, C. (1997) The racial origins of zoning in American cities, in *Urban Planning and the African American Community: In the Shadows* (eds J.M. Thomas and M. Ritzdorf), Sage Publications, Inc., Thousand Oaks, pp. 23–42.

Shanahan, D.F., Fuller, R.A., Bush, R. *et al.* (2015) The health benefits of urban nature: how much de we need? *BioScience*, **65**, 476–485.

Shanahan, D.F., Lin, B.B., Gaston, K.J. *et al.* (2011) Socio-economic inequalities in access to nature on public and private lands: a case study from Brisbane, Australia. *Landscape and Urban Planning*, **130**, 14–23.

Shin, H.B. and Li, B. (2012) Migrants, landlords and their uneven experiences of the Beijing Olympic Games, in Centre for Analysis of Social Exclusion Paper *163*, London School of Economics, London.

Smith, K.R., Brown, B.B., Yamada, I. *et al.* (2008) Walkability and body mass index: density, design, and new diversity measures. *American Journal of Preventive Medicine*, **35**, 237–244.

Soule, D. (ed). (2009) *Urban Sprawl: A Comprehensive Reference Guide*. Greenwood Press, Westport.

Spencer, R.D., Beausoleil, R.A., and Martorello, D.A. (2007) How agencies respond to human-black bear conflicts: a survey of wildlife agencies in North America. *Ursus*, **18**, 217–229.

Steinbrink, M. (2013) Festifavelisation: mega-events, slums and strategic city-staging – the example of Rio de Janeiro. *Journal of the Geographical Society of Berlin*, **144**, 129–145.

Steinemann, A.C. (2009) Fragranced consumer products and undisclosed ingredients. *Environmental Impact Assessment Review*, **29**, 32–38.

Steinemann, A.C., MacGregor, I.C., Gordon, S.M. *et al.* (2011) Fragranced consumer products: chemicals emitted, ingredients unlisted. *Environmental Impact Assessment Review*, **31**, 328–333.

Sugiyama, T., Francis, J., Middleton, N.J. *et al.* (2010) Associations between recreational walking and attractiveness, size, and proximity of neighborhood open spaces. *American Journal of Public Health*, **100**, 1752–1757.

Sugiyama, T., Neuhaus, M., Cole, R. *et al.* (2012) Destination and route attributes associated with adults' walking: a review. *Medicine and Science in Sports and Exercise*, **44**, 1275–1286.

Sullivan, B.K., Kwiatkowski, M.A., and Schuett, G.W. (2004) Translocation of urban Gila Monsters: a problematic conservation tool. *Biological Conservation*, **117**, 235–242.

Tan, J., Zheng, Y., Tang, X. *et al.* (2010) The urban heat island and its impact on heat waves and human health in Shanghai. *International Journal of Biometeorology*, **54**, 75–84.

The Editors (1972) Introductory statement. *Human Ecology*, **1**, 1.

Thornton, C. and Quinn, M.S. (2009) Coexisting with cougars: public perceptions, attitudes, and awareness of cougars on the urban-rural fringe of Calgary, Alberta, Canada. *Human-Wildlife Conflicts*, **3**, 282–295.

Tibaijuka, A.K. (2005) *Report of the Fact-finding Mission to Zimbabwe to Assess the Scope and Impact of Operation Murambatsvina by the UN Special Envoy on Human Settlement Issues in Zimbabwe*. UN-Habitat, Nairobi. http://www1.umn.edu/humanrts/research/ZIM%20UN%20Special%20Env%20Report.pdf (accessed 12/06/2015)

Turner, W.R., Nakamura, T., and Dinetti, M. (2004) Global urbanization and the separation of humans from nature. *BioScience*, **54**, 585–590.

Ulrich, R.S. (1984) Human response to vegetation and landscapes. *Landscapes and Urban Planning*, **13**, 29–44.

UN-Habitat. (2003) *Slums of the World: The Face of Urban Poverty in the New Millennium*? UN-Habitat, Nairobi. http://www.sustainable-design.ie/sustain/UN-Habitat_2003WorldSlumsReport.pdf (accessed 12/06/2015)

Valavanidis, A., Fiotakis, K., and Vlachogianni, T. (2008) Airborne particulate matter and human health: toxicological assessment and importance of size and composition of particles for oxidative damage and carcinogenic mechanisms. *Journal of Environmental Science and Health, Part C: Environmental Carcinogenesis and Ecotoxicology Reviews*, **26**, 339–362.

Veitch, J., Salmon, J., Carver, A. *et al.* (2014) A natural experiment to examine the impact of park renewal on park-use and park-based physical activity in a disadvantaged neighbourhood: the REVAMP study methods. *BMC Public Health*, **14**, 1–9.

Viegi, G., Simoni, M., Scognamiglio, A. *et al.* (2004) Indoor air pollution and airway disease. *International Journal of Tuberculosis and Lung Disease*, **8**, 1401–1415.

Wang, X.Y., Barnett, A.G., Hu, W. *et al.* (2009) Temperature variation and emergency hospital admissions for stroke in Brisbane, Australia, 1996-2005. *International Journal of Biometeorology*, **53**, 535–541.

WAPC. (2009) *Liveable Neighbourhoods: A Western Australian Government Sustainable Cities Initiative*. Western Australian Planning Commission, Perth. http://www.planning.wa.gov.au/dop_pub_pdf/LN_Text_update_02.pdf (accessed 12/06/2015)

Weeks, J.R., Hill, A., Stow, D. *et al.* (2007) Can we spot a neighborhood from the air? Defining neighborhood structure in Accra, Ghana. *GeoJournal*, **69**, 9–22.

Weiser, S.D., Leiter, K., Bangsberg, D.R. *et al.* (2007) Food insufficiency is associated with high-risk sexual behaviour among women in Botswana and Swaziland. *PLoS Medicine*, **4**, 1589–1598.

Weng, Q. and Yang, S. (2004) Managing the adverse thermal effects of urban development in a densely populated Chinese city. *Journal of Environmental Management*, **70**, 145–156.

White, L.A. and Gehrt, S.D. (2009) Coyote attacks on humans in the United States and Canada. *Human Dimensions of Wildlife*, **14**, 419–432.

World Bank. (2013) *Global Monitoring Report 2013: Rural-Urban Dynamics and the Millennium Development Goals*. World Bank, Washington DC. doi: 10.1596/978-0-8213-9806-7

World Health Organization. (2011) *Intersectoral Actions in Response to Cholera in Zimbabwe: From Emergency Response to Institution Building*. World Conference on Social Determinants of Health, Rio de Janeiro, Brazil, Draft Background Paper no. 23. http://www.who.int/sdhconference/resources/draft_background_paper23_zimbabwe.pdf (accessed 12/06/2015)

World Health Organization and UN-Habitat (2010) *Hidden Cities: Unmasking and Overcoming Health Inequities in Urban Settings*, WHO Press, Geneva.

Wyly, E.K. and Hammel, D.J. (1999) Islands of decay in seas of renewal: housing policy and the resurgence of gentrification. *Housing Policy Debate*, **10**, 711–771.

Zhang, J. and Smith, K.R. (2003) Indoor air pollution: a global health concern. *British Medical Bulletin*, **68**, 209–225.

Zheng, H.W., Shen, G.Q., and Wang, H. (2014) A review of recent studies on sustainable urban renewal. *Habitat International*, **41**, 272–279.

Zhou, Z., Dionisio, K.L., Arku, R.E. *et al.* (2011) Household and community poverty, biomass use, and air pollution in Accra, Ghana. *Proceedings of the National Academy of Sciences of the United States of America*, **108**, 11028–11033.

Zhu, P. and Zhang, Y. (2008) Demand for urban forests in United States cities. *Landscape and Urban Planning*, **84**, 293–300.

Zulu, E.M., Dodoo, F.N., and Chika-Ezeh, A. (2002) Sexual risk taking behaviour in the informal settlements of Nairobi, Kenya, 1993-1998. *Population Studies*, **56**, 311–323.

CHAPTER 7

Conserving biodiversity and maintaining ecosystem services in cities

7.1 Introduction

As outlined in earlier chapters, the construction and expansion of cities can have substantial adverse impacts on native biodiversity, and many species around the world are currently threatened by urbanization (McKinney 2002; Pauchard et al. 2006; McDonald et al. 2008; Goddard et al. 2010). However, this should not lead us to regard cities as unimportant places for biological conservation. Urban, suburban and peri-urban areas still support a vast array of native species and ecological communities, from the critically endangered to those currently of least conservation concern (Yencken and Wilkinson 2000; Alvey 2006; Aronson et al. 2014a; Ives et al. 2016; Box 7.1). If we were to focus conservation efforts only on rural and wilderness areas, many of these species and communities would be lost. In addition to its intrinsic value, biodiversity in cities has important benefits for the health and wellbeing of urban-dwelling humans (See Chapter 6; Dearborn and Kark 2010; Secretariat of the Convention on Biological Diversity 2012). A substantial body of knowledge from the fields of urban ecology, population health, epidemiology and environmental psychology demonstrates that urban biodiversity increases the liveability of cities for humans (Fuller et al. 2007; Tzoulas et al. 2007; Taylor and Hochuli 2015; Carrus et al. 2015).

Ecosystem functions are functions performed by ecosystems, such as nutrient cycling, pollination, primary and secondary production, the regulation of climate and the absorption of pollutants (see Chapter 5). Ecosystem services are benefits that humans derive from ecosystem functions – these include plant and animal foods, timber, plant fibres, clean air, clean water and a more comfortable climate (de Groot et al. 2012). This list obviously includes many things that are vital for human survival. The total global value of ecosystem services is enormous – estimated at US$125 trillion/year in 2011 (Costanza et al. 2014). Considering individual habitat types, the value of the ecosystem services provided by coral reefs, wetlands and tropical forests has been estimated at $352,249, $140,174 and $5382 per hectare per year, respectively. In this same analysis, the

Ecology of Urban Environments, First Edition. Kirsten M. Parris.

Box 7.1 Urban Avoiders, Adapters and eExploiters

McKinney (2002) first used the terms urban avoiders, adapters and exploiters to describe three groups of species that respond to urbanization in different ways. He defined urban avoiders as species that are very sensitive to human persecution and habitat disturbance, such as large-bodied carnivores, ground-nesting birds, and late-successional plants. He considered urban adapters to be species that can persist successfully in suburban and urban-fringe habitats by utilizing available niches and resources. This group includes guilds such as omnivorous and ground-foraging birds, burrowing mammals, medium-sized predators, early-successional plants and bird-dispersed shrubs (McKinney 2002). Urban exploiters (also known as synanthropes or commensal species) are a small group of species that show a strong positive response to the resources provided by humans in urban environments, and attain very high densities in cities around the world. This group includes animals such as the rock dove *Columba livia*, European starling *Sturnus vulgarus*, house mouse *Mus musculus*, black rat *Rattus rattus*, cockroach (various species), and a diversity of ruderal plants with wind-dispersed seeds and a high tolerance of disturbance (McKinney 2002).

While McKinney's (2002) classification is a useful one, the terms "avoider", "adapter" and "exploiter" imply a particular intent or agency on behalf of these species that may not be appropriate. For example, urban avoiders may not actively choose to avoid urban habitats. Instead, populations of these species are not persisting in urban environments for a range of other reasons (e.g., because they have particular traits that make them vulnerable to disturbance or predation in cities, or resource requirements that cannot easily be met in urban landscapes). Similarly, urban adapters do not necessarily change their behaviour or resource requirements to match what is available in urban environments. Grant et al. (2011) identified three groups of amphibians and reptiles on the basis of their response to urbanization – urbanophiles, urbanophobes (after Mitchell et al. 2008) and the urbanoblivious. The first two groups roughly correspond to McKinney's urban exploiters and urban avoiders. The third is comprised of cryptic species that can persist for many years in small, isolated pockets of habitat surrounded by urban development, largely unnoticed by humans. North American examples of the urbanoblivious include long-lived, large-bodied species such as the snapping turtle *Chelydra serpentina*, as well as short-lived, small-bodied species such as the red-backed salamander *Plethodon cinereus* (Grant et al. 2011).

Particular ecological and life-history traits of species are clearly associated with their persistence in, or exclusion from, urban habitats. The nature and generality of these traits are currently attracting considerable interest in urban-ecological research (e.g., Williams et al. 2005; Croci et al. 2008; Knapp et al. 2008; Hahs et al. 2009; Grant et al. 2011; see Chapter 4 for further discussion). It has been suggested that species tolerant of city life occupy a broader niche than urban avoiders, giving them greater behavioural, physiological and ecological flexibility to persist in altered urban environments (Bonier et al. 2007). This pattern appears to hold for birds, with species common in urban environments having a broader latitudinal and elevational distribution than congeneric species that are more abundant in rural environments (Bonier et al. 2007). It also appears to hold for amphibians and reptiles, with many species that are common in cities demonstrating broad habitat and dietary tolerances (Mitchell et al. 2008; Grant et al. 2011). Conservation of some urban avoiders/urbanophobes in cities may not be possible, but in other cases, targeted conservation actions may enable some of these species to coexist with humans.

ecosystem services delivered by urban habitats were valued at $6661 per hectare per year in 2011 (Costanza et al. 2014). Important services provided by urban ecosystems include nutrient cycling, shading, cooling, absorption of atmospheric pollutants, carbon sequestration, primary and secondary production (including urban agriculture), hydrological cycling, biological control of pest species and the treatment of waste (Jo 2002; Li et al. 2005; Dwivedi et al. 2009). Biodiversity is known to enhance certain ecosystem services in cities, including nutrient cycling, microclimatic cooling, absorption of atmospheric pollutants and hydrological cycling (Bolund and Hunhammer 1999; MEA, 2005).

The conservation of biodiversity and the provision of ecosystem services in cities can be challenging, with many competing interests in operation (Eppink et al. 2004; Polasky et al. 2008; Gordon et al. 2009; UN-HABITAT 2010; Bekessy et al. 2012). As land is often in short supply, prices are high. This limits opportunities to purchase land for new parks or conservation reserves within or close to cities (Snyder et al. 2007), except in places that are unsuitable for other sorts of development (e.g., disused industrial or landfill sites, areas of steep terrain or areas subject to flooding). The high premium on land also means that remnant, biodiverse habitats in cities – such as wetlands, riparian corridors and patches of remnant vegetation – are under constant threat from urban development (Bekessy and Gordon 2007). In many cases, short-term economic interests or other socioeconomic considerations override the long-term benefits of conserving biodiversity and ecosystem services (Crook and Clapp 1998; Balmford et al. 2002; Breed et al. 2014). Thirdly, the constant outward march of sprawling cities means that biodiverse areas, currently some distance from the edge of a city, may eventually be overrun by suburbia or informal settlements (Polk et al. 2005; Benítez et al. 2012; Seto et al. 2012). The ecosystem services provided by trees, grasses, animals, fungi, microbes, streams, wetlands and permeable soils in cities are often unrecognised or undervalued (Taylor and Hochuli 2015), making it easy for these organisms and habitat elements to be altered or destroyed. In this chapter, I present some practical strategies for the conservation of biodiversity and the provision of ecosystem services in urban environments – for the benefit of humans and all the other species that live in cities.

7.2 Strategies for conserving biodiversity and maintaining ecosystem services in cities

7.2.1 Integrate urban ecology with urban planning and design

Effective conservation of biodiversity and ecosystem services in urban environments will require input from a wide range of disciplines, including ecology, urban planning, landscape architecture, ecological economics, engineering and landscape management (McPherson et al. 1997; Ong 2003; Hunter and Hunter 2008; Ahern 2013; Tanner et al. 2014). Planners, designers and landscape architects have an important role to play in protecting and promoting biodiversity, green infrastructure and ecosystem services in cities, as their practice has a direct and lasting influence on the urban form (Breed et al. 2014; Hes and Du Plessis 2015). Many opportunities exist to improve the conservation of biodiversity and the provision of ecosystem services in cities by better integrating urban ecology into urban planning and landscape management across a variety of spatial scales, from individual house lots to neighbourhoods, precincts and regions (e.g., Jim and Chen 2003; Uy and Nakagoshi 2008; Gordon et al. 2009; Cilliers et al. 2014; Garrard and Bekessy 2014).

Conservation planning can be integrated into broad-scale urban planning via strategic assessments of the possible environmental impacts associated with urban development. Strategic environmental assessments are regional-scale assessments of the impact of proposed developments on matters of environmental significance, including threatened species, threatened ecological communities and natural sites of biological, social and/or economic significance. Strategic assessments can incorporate environmental values and goals into broad-scale policy and planning at an early stage, consider alternative development scenarios, and assess environmental impacts at a landscape scale (Chaker et al. 2006). Strategic environmental assessments may provide better protection for biodiversity and ecosystem services in cities than smaller-scale environmental impact assessments, which often lack a long-term vision, a broad-scale plan and a mechanism for considering the cumulative impacts of many small actions. These cumulative impacts have been described as "death by a thousand cuts"; each proposed development is considered independently of other developments in the region and its likely environmental impact is not sufficiently large to halt proposed works. But when taken in combination, these many small actions have a significant impact on the persistence of threatened species and ecological communities. On the other hand, if strategic environmental assessments are not sufficiently rigorous, their regional-scale approach may actually accelerate the loss of native species and ecological communities from urban areas by facilitating the broad-scale and rapid clearing of habitat (Parris 2009).

For example, State and Territory governments in Australia have prepared strategic environmental assessments under the Commonwealth *Environmental Protection and Biodiversity Conservation Act 1999* to gain federal approval for proposed programs of urban development in Melbourne, Sydney and Canberra (DSE 2009; NSW Department of Planning 2010; Umwelt 2013). The proposed expansion of Melbourne's urban growth boundary in 2009 included 40,000 ha of new housing, a regional rail link and a ring road to be built within four growth corridors around the city. These corridors coincided with areas supporting populations of federally listed threatened species, such as the growling grass frog (Figure 7.1), striped legless lizard, golden sun moth, southern brown bandicoot, spiny rice-flower and matted flax-lily, and remnants of critically endangered vegetation communities (DSE 2009). The draft strategic impact assessment report indicated that the

Figure 7.1 A growling grass frog *Litoria raniformis*. This species is threatened by the continued expansion of Melbourne, Australia. Photograph by Geoff Heard.

proposed development would destroy around 8000 ha (80 km^2) of critically endangered natural temperate grassland, grassy eucalypt woodland and lowland plains grassy wetland and have substantial (but unquantified) impacts on a range of threatened species (DSE 2009). In addition, the draft report only presented one option for development and did not assess the impacts of the proposed development on landscape function and connectivity. Thus, it did not meet all the criteria required for a robust strategic impact assessment (IAIA 2002; Chaker et al. 2006). Later iterations of the process reduced the area of endangered vegetation communities to be cleared to around 4500 ha.

The Victorian Government is in the process of establishing two new grassland reserves in Melbourne's west through compulsory acquisition of private land, in an effort to offset the grassland lost to urban development (DSE 2011). And as part of the strategic impact assessment process, it has prepared a Biodiversity Conservation Strategy and a series of Sub-regional Species Strategies for threatened species, including the growling grass frog *Litoria raniformis* (DEPI 2013a, b). This species was once widely distributed throughout south-eastern Australia (Pyke 2002), but population declines and local extinctions since the late 1970s mean that it now occupies only small pockets of its original range (Mahony 1999). Remnant populations around Melbourne are threatened by the loss, degradation and fragmentation of habitat resulting from urban expansion (Heard et al. 2012); the current distribution of *L. raniformis* in greater Melbourne substantially overlaps the new growth corridors to the north, west and south-east of the city

(DEPI 2013b). However, in a practical example of the integration of urban ecology and urban planning, a stochastic metapopulation model for *L. raniformis* (Heard et al. 2013) is currently being used to predict the impact of different development and management scenarios on the species prior to development. These scenarios include the protection of existing wetlands, the retention of buffer zones of various widths along streams, and the creation of new wetlands (Heard et al. 2013). The model can identify the optimal solutions and the optimal boundaries of conservation areas to best protect growling grass frogs with a given budget.

7.2.2 Protect biodiverse landscape features and important biophysical assets

Patches of remnant native vegetation, secondary vegetation, ephemeral and permanent wetlands, streams and drainage lines often support a high diversity of species that are not found elsewhere within the urban landscape, while also providing important ecosystems services including nutrient and hydrological cycling, pollination, carbon sequestration and flood mitigation (Dearborn and Kark 2010; Threlfall et al. 2012; Calhoun et al. 2014; Diaz-Porras et al. 2014). Although they may be small in area, the ecological value of these features is disproportionately large; as a consequence, they are sometimes referred to as keystone features or structures (Stagoll et al. 2012; Calhoun et al. 2014). Conserving these features will therefore provide a substantial benefit to urban biodiversity per unit area. For example, large, old trees act as keystone structures in urban environments, supporting a high diversity of invertebrates, birds, bats and arboreal mammals, while providing shade, microclimatic cooling, carbon sequestration and important habitat features, such as hollows (cavities) and coarse woody debris (Carpaneto et al. 2010; Lindenmayer et al. 2012; Stagoll et al. 2012; Diaz-Porras et al. 2014).

The species richness, abundance and breeding activity of birds increased with increasing availability of large trees in urban parks in Canberra, Australia, during a recent study; the largest trees (>100 cm DBH (diameter at breast height)) demonstrated the greatest habitat-provision benefits (Stagoll et al. 2012). Saproxylic scarab beetles, including the IUCN-listed hermit beetle *Osmoderma eremita*, were more likely to be present in tall, live, hollow-bearing trees in the Villa Borghese Park in Rome, Italy-particularly those trees with abundant wood mould in their hollows (Carpaneto et al. 2010). However, many of these same trees were considered a danger to humans and were scheduled for removal, highlighting a conflict between biodiversity conservation and concern for public safety in cities (Carpaneto et al. 2010; Le Roux et al. 2014). A similar conflict contributes to a scarcity of fallen logs – another key habitat element for many species – in urban environments. In this case, logs are considered hazardous because of a (largely unsupported) perception that they provide fuel for wildfires (Lehvävirta 2007; Le Roux et al. 2014). The risk posed to humans by large, old trees in urban environments could be managed through fencing around trees or landscaping to separate them from the high-use areas of a park, such as playgrounds and walking paths (Stagoll et al. 2012). Educating the public about the actual fire risk posed by fallen logs of large diameter may increase their social acceptance in urban parks.

The preservation of upland drainage lines (ephemeral, first-order streams) in urban environments would allow natural drainage following rainfall events and help to reduce the volume of stormwater entering streams and other receiving waters (Levick et al. 2008). Linear parks set aside around drainage lines could also provide open green space for humans and habitat for urban biodiversity (Uy and Nakagoshi 2008; Maric et al. 2013). However, because these drainage lines only contain flowing water after substantial rainfall events, they are often built over and their natural drainage function replaced with engineered structures such as pipes. Two successful examples of preserved urban drainage lines are the natural drainage system of the Village Homes subdivision in Davis, California – a network of creeks, swales and ponds that retains rainwater and stores moisture in the soil (Corbett and Corbett 2000; Karvonen 2011) – and the Kronsberg Neighbourhood near Hannover, Germany (Rumming 2004; Coates 2013). These kinds of design practices reduce the degradation of urban catchments and downstream waters, provide green space for human recreation and habitat connectivity, and promote biodiversity in the wider urban landscape (Urbonas and Doerfer 2005; Walsh et al. 2005; von Haaren and Reich 2006).

Small patches of remnant vegetation in cities can continue to support a surprisingly high diversity of native species, even though a city has grown up around them (Williams et al. 2006; Gupta et al. 2008; Newbound et al. 2012). For example, a survey of fungi in patches of remnant woodland in Melbourne, Australia, found between 54 and 114 species per patch, using analysis of fungal DNA from soil samples (Newbound et al. 2012), while populations of at least 135 vascular plant species were still persisting in an urban forest remnant in Nagpur, India (Gupta et al. 2008). Semi-constructed habitat features such as golf courses and recreational parks can also be important spaces for biodiversity and significant providers of ecosystem services in cities, particularly if some areas of natural vegetation or ground cover are maintained (e.g., leaf litter and/or unmown grass; Savard et al. 2000; Barthel et al. 2005; Tyrvainen et al. 2005; Yasuda and Koike 2006; Colding and Folke 2008; Stagoll et al. 2012). A study of golf courses, urban parks and residential gardens in Melbourne, Australia found that golf courses supported a greater diversity and abundance of beetles, native bees and bugs (order Heteroptera) – including predatory bugs that are important for the control of insect pests – than nearby parks and gardens (Threlfall et al. 2014). Golf courses also hosted more species of birds and bats and supported almost twice the breeding activity of birds as the other two habitat types. The importance of golf courses as havens for biodiversity and providers of ecosystem services is likely to increase as the surrounding land experiences greater anthropogenic impacts (Colding and Folke 2008). Old gravel pits are another semi-constructed habitat feature that may be important for the conservation of invertebrates in urban environments, including butterflies (Lenda et al. 2012).

Private gardens and street trees form extensive living assets in cities, and comprise a substantial proportion of the total cover of trees and permeable soils in urban landscapes (Cameron et al. 2012). Consequently, they play an important role in the provision of ecosystem services, such as shading, microclimatic cooling, nutrient cycling, primary production, carbon sequestration, noise reduction, absorption of atmospheric pollutants, hydrological cycling and a sense of connectedness to nature (Livesley et al. 2010; Kendal et al. 2012; Norton et al. 2013; Clarke et al. 2014; Stovin et al. 2015). Private gardens also support a significant proportion of urban biodiversity across many taxonomic groups, from soil microbes and fungi to ornamental, edible and medicinal plants, bugs, bees, butterflies, reptiles and birds (Gaston et al. 2005; Loram et al. 2008; Akinnifesi et al. 2010; Goddard et al. 2010; Kendal et al. 2010; Jaganmohan et al. 2012; Cilliers and Siebert 2012). However, trends towards urban infill and the construction of larger houses mean that the proportional cover of private gardens (and thus the ecosystem services and biodiversity they support) is declining in many cities (e.g., Marriage 2010; Vallance et al. 2012; Brunner and Cozen 2012).

In some parts of the world, there is a trend towards larger houses that take up a much higher proportion of the lot than was customary only a few decades ago, leaving a smaller area available for garden (e.g., Marriage 2010; Hall 2011). In 2009, Australia had the largest new, free-standing houses in the world, with an average floor area of 245.3 m^2 – an increase of 34% over 20 years (James 2009). Based on US census data, the average floor area of new, single-family houses in the USA was 241.4 m^2 in 2013, compared with 154.2 m^2 in 1973 (Figure 7.2; US Census Bureau 2014). A similar trend can be observed in New Zealand, where the average floor area of new houses was 219 m^2 in 2011, compared with 139 m^2 in 1991 (Marriage 2010; Fig. 7.2). A high ratio of house to garden on a lot decreases amenity; the benefits of shading by trees and microclimatic cooling provided by transpiring green vegetation are lost, the cover of heat-absorbing impermeable surfaces increases, and there is less space for outdoor play and recreation. By making the local climate hotter and drier, this urban form increases the use of air-conditioning in summer and overall energy consumption (and in many cases, the combustion of fossil fuels). It also decreases the volume of rainwater absorbed by soils and increases stormwater runoff to streams (Brunner and Cozen 2012).

7.2.3 Grow the green city

Given that street trees and gardens are so important for biodiversity and the liveability of cities – and are likely to become more so under climate change (Gago et al. 2013; Norton et al. 2013) – cities need visionary urban greening policies to increase tree cover, as well as stronger regulation to protect existing vegetation

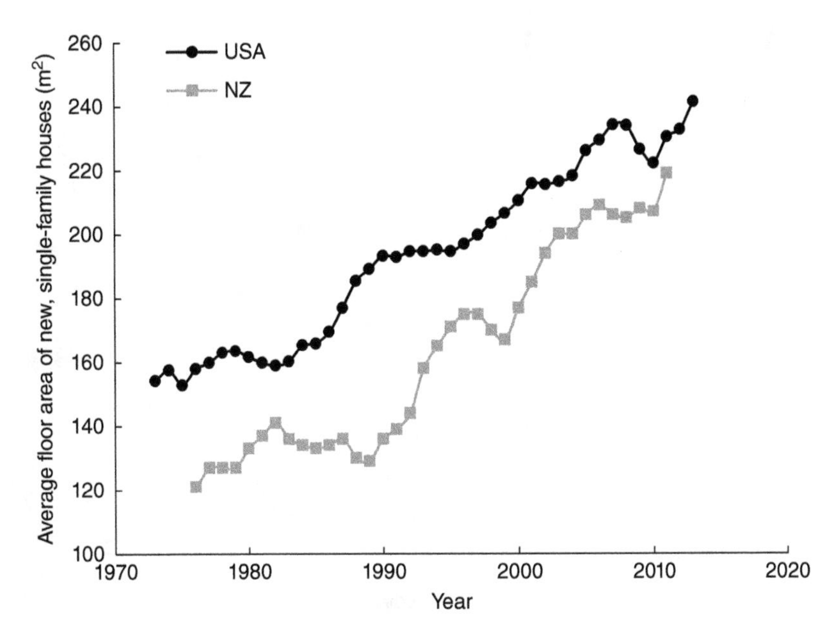

Figure 7.2 Average floor area of new, single-family houses in the USA and NZ versus year; in both countries, house sizes have increased substantially over recent decades. Data from Marriage (2010) and US Census Bureau (2014).

and permeable soils. These kinds of regulations are already in place in a number of cities, including Berlin (the Biotope Area Factor) and Seattle (the Seattle Green Factor). A tree-protection policy in the Woodlands Township, Texas sets a minimum cover of trees and shrubs and strictly regulates the removal of trees, with a permit required to cut down any tree with a diameter ≥15 cm (Sung 2013). This policy is effective in reducing the urban heat-island effect, with average land surface temperatures 1.5–3.9°C lower than those of nearby control neighbourhoods without a tree-protection policy. The cooling effect of the extra vegetation in The Woodlands Township is more pronounced in summer when heat mitigation is most needed, with an average temperature differential of up to 5°C and a largest observed differential of 7.1°C (Sung 2013). Over the seven neighbourhoods included in this study, land surface temperature on a hot day in May decreased with increasing tree cover (Figure 7.3).

As discussed in Chapter 6, a variety of studies have shown that the cover of trees and other leafy vegetation in cities is positively correlated with the socio-economic status of established neighbourhoods; in many cities, wealthier areas have more trees (Pedlowski et al. 2002; Escobedo 2006; Harlan et al. 2007; Weeks et al. 2007; Clarke et al. 2013; Shanahan et al. 2014; but see Hetrick et al. 2013). This uneven distribution of urban trees and other vegetation across cities means that the community benefits and ecosystem services provided by urban tree cover are also

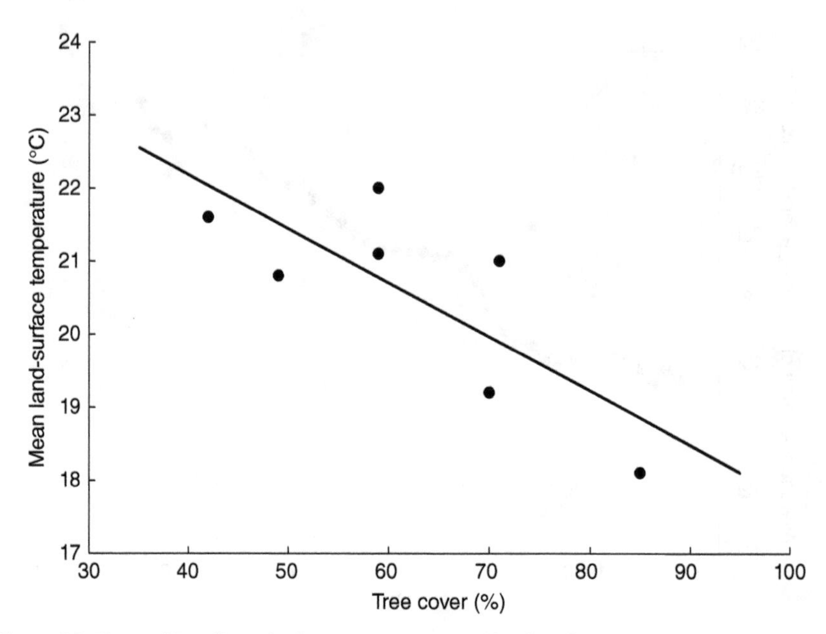

Figure 7.3 The cooling effect of urban tree cover. Mean land surface temperature (LST) versus tree cover (%) across each of seven neighbourhoods in the northern Houston metropolitan area, Texas, USA. Temperature data derived from thermal remote sensing, averaged across 37 Landsat TM images taken between 2000 and 2010; tree cover data derived from the US national agricultural imagery program (NAIP) aerial photography taken on May 3, 2010. Data from Sung 2013.

unevenly distributed. Urban tree cover can therefore be considered as a social justice issue (Wolch et al. 2014). Given that a 5% reduction in tree cover can translate to a 1–2°C increase in air temperature (McPherson and Rowntree 1993), differences in tree cover between wealthy and poorer suburbs can create substantial differences in temperature between the most- and least-treed areas of a city. To improve the equity of urban tree cover, urban greening policies and tree planting strategies should be targeted to neighbourhoods with a low cover of vegetation and a human population that is vulnerable to extreme heat (Norton et al. 2015). These strategies could include incentive schemes for householders to increase the cover of trees and shrubby vegetation in their own garden, as well as planting more trees in streets and public parks (Shanahan et al. 2014). In hot and dry climates, stored stormwater could be used to irrigate parks, gardens and street trees, enhancing urban cooling and improving human thermal comfort (Coutts et al. 2013). This would have the additional benefit of reducing stormwater runoff to local streams and other receiving waters.

7.2.4 Maintain or re-establish landscape connectivity

Habitat features in cities, such as parklands, wetlands and patches of remnant or secondary vegetation, are often separated from each other by expanses of roads, buildings and other urban infrastructure. As discussed in Chapter 3, this can impede or prevent the successful movement of propagules and individuals between discrete areas of habitat, effectively isolating populations of many fungi, plants and animals (Beier 1995; Bjurlin et al. 2005; Parris 2006; Fu et al. 2010; Hale et al. 2013; Ward et al. 2015). To avert this isolation, landscape connectivity needs to be considered during urban planning – preferably before new cities or suburbs are developed. Preserving undeveloped linear features to act as corridors or conduits between biodiverse landscape features will help to maintain connectivity between populations for some taxonomic groups in urban environments, including arthropods, amphibians and mammals (Beier and Noss, 1998; Indykiewicz et al. 2011; Vergnes et al. 2012, 2013; Hale et al. 2013; Wilson et al. 2013). In Paris, France, wooded corridors are used by arthropods to travel through the urban landscape; private gardens connected to woodlands via these corridors (termed connected gardens) were found to host arthropod communities that were functionally similar to those of the woodlands (Vergnes et al. 2012). Connected gardens also had a higher abundance of spiders and rove beetles (staphylinids) than disconnected gardens. A second study found that these same corridors are used by shrews to move from woodlands to private gardens, with shrews observed in 43% of connected gardens and only 6% of disconnected gardens (Vergnes et al. 2013).

An urban green space plan for Hanoi, Vietnam, includes parks, large green wedges, one green belt and a number of other linear green features along roadsides and streams, which together form an ecologically effective green network (Figure 7.4; Uy and Nakagoshi 2008). This network has been designed to connect different areas of green space and allow the movement of fauna and flora between them. It contains elements of existing green space as well as plans for additional greenways to be constructed in the future, and incorporates both natural and constructed linear features into the green network. The Hanoi green space plan demonstrates the possibility of retrofitting established urban areas to enhance habitat connectivity; however, it is easier and possibly more economical to preserve existing landscape connections than to create new ones (Benedict and McMahon 2006). Given that habitat corridors and other landscape connections occupy physical space, opportunities to recreate old corridors or establish new ones may be limited in densely developed urban environments. Underpasses, overpasses, wildlife bridges and other engineered features that allow animals to cross roads may also make a valuable contribution to the provision of habitat connectivity in cities if certain conditions are met; the target animals must actually use the structures in sufficient numbers and remain safe from predators as they do so (Ng et al. 2004; van der Grift et al. 2013; Hamer et al. 2014; Ward et al. 2015).

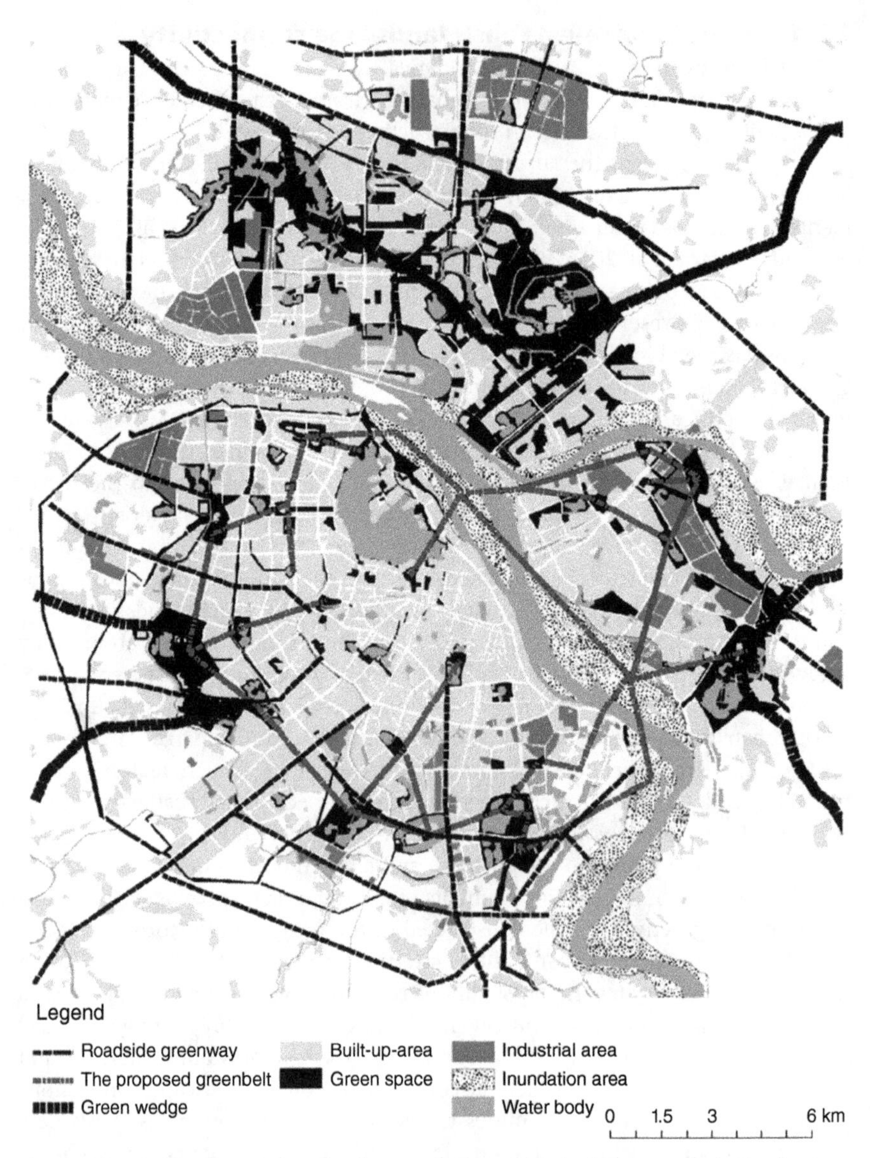

Figure 7.4 An urban green space plan for Hanoi, Vietnam, including an ecologically effective network of parks, green wedges and linear green features along roadsides and streams. Uy & Nakagoshi 2008, Figure 10. Reproduced with permission of Elsevier.

Humans can personally assist the movement of animals and plants across the urban landscape, either by moving (translocating) individual plants or animals from one location to another, or by helping migrating animals to arrive safely at their destination. While translocation is often proposed as a suitable tool for species conservation in urban habitats (e.g., moving individuals from a site that is about to be developed to another, apparently suitable site), it generally has a low probability of success. Many translocations of reptiles, amphibians and mammals fail for a variety of reasons, including the homing of released animals, a shortage of suitable, unoccupied habitat or a decline in habitat suitability over time at the release site, an insufficient number of individuals released, predation and/or disease (Pietsch 1994; Fischer and Lindenmayer 2000; Armstrong 2008; Germano and Bishop 2009; Koehler and Gilmore 2015). Although intuitively appealing, translocation is no panacea for the destruction of important habitat features in the urban landscape, such as wetlands or patches of remnant vegetation. In some circumstances, however, translocation may be appropriate for assisting the recolonization of suitable but unoccupied habitat in cities that is too isolated to be recolonized naturally (Martell et al. 2002; Parris 2006; White and Pyke 2008).

Animals that migrate seasonally from one part of their habitat to another often follow the same routes to the same destinations year after year, ignoring changes in the landscape. For example, frogs may arrive to breed at a wetland that no longer exists (Windmiller and Calhoun 2008). Many thousands of frogs, salamanders, newts and turtles die each year attempting to cross roads on their way to or from breeding sites (Gibbs and Shriver 2002, 2005; Marchand and Litvaitis 2004; Windmiller and Calhoun 2008). In a variety of countries, including South Africa, the UK, The Netherlands, Romania, Germany, Denmark and Estonia, community groups have mobilized to help prevent the road kill of amphibians through a practice known as patrolling. Volunteers patrol important migratory road crossings during the appropriate season and help the animals to cross the road safely. Froglife in the UK has a "Toads on Roads" programme for the common toad *Bufo bufo*, which allows members of the community to register a toad crossing and volunteer as a toad patroller (www.froglife.org). Similar programmes operate in Cape Town, South Africa, to help endangered western leopard toads *Amietophrynus pantherinus* get to and from their breeding sites safely. The Toad NUTS (Noordhoek unpaid toad savers) group recently established a drift fence with pitfall traps along a busy road in Noordhoek (a suburb in Cape Town) to intercept moving toads so they could then be carried safely to the other side. There are also examples of reduced speed limits along roads frequented by amphibians, road signs to warn motorists and even seasonal road closures to protect migrating animals (Hilty et al. 2006; Wang et al. 2013; Figure 7.5).

7.2.5 Use small spaces

The urban environment can be very crowded with human-made structures, paved surfaces, and a fine-scale mosaic of land tenures. This can make it difficult to find space for natural areas and the vegetation, permeable soils, wetlands

Figure 7.5 A sign in Angelbachtal, Germany to warn motorists that frogs may be crossing the road. Picture has been straightened, cropped and converted to black and white. Photograph by Jencu. https://www.flickr.com/photos/jennycu. Used under CC-BY-2.0 https://creativecommons.org/licenses/by/2.0/

and other landscape features that act as habitat for biodiversity and providers of ecosystem services. One solution to this problem is to use small spaces throughout the city – horizontal, vertical, and everything in between (Rosenzweig 2003a, b). Reconciliation ecology is a branch of ecology that emphasizes the conservation of biodiversity in human-dominated landscapes, including urban landscapes. First introduced in 2003, reconciliation ecology can be defined as "the science of inventing, establishing and maintaining new habitats to conserve species diversity in places where people live, work or play" (Rosenzweig 2003a). One focus of reconciliation ecology is small-scale or local changes to urban environments that can help to conserve and enhance biodiversity, on both public and private land. The practice of wildlife gardening is a form of reconciliation ecology, and an effective way to increase the biodiversity of cities across many small, private gardens (Rudd et al. 2002; Colding 2007; Gaston et al. 2007; Goddard et al. 2013).

A large city may contain many thousands of flat-roofed buildings that could be retrofitted to install living, green roofs. Green roofs can provide a range of benefits, including insulation and thermal regulation of the buildings upon which they grow, cycling of stormwater via absorption by the growing medium and evapotranspiration by plants, habitat for a diversity of taxonomic groups, removal of pollutants from the urban atmosphere, and improved productivity and mental health among people who can observe a green roof throughout their work day (Kadas 2002; Wong et al. 2003; Mentens et al. 2006; Oberndorfer et al. 2007; Currie and Bass 2008; Castleton et al. 2010; Braaker et al. 2014; Lee et al. 2015). Roof-top gardens can be cultivated to provide a pleasant space for relaxation, to grow herbs, fruit and vegetables, and even to raise bees for honey; in so doing, they provide food, amenity, a connection to nature and a variety of other ecosystem services including pollination (Shariful Islam 2004; Wong et al. 2003; Grewal and Grewal 2012). Urban rooftop beekeeping is thriving in many cities around the world; London leads the way with more than 3000 rooftop hives, while rooftop beekeepers can also be found in Shanghai, Honolulu, Brisbane and New York (Saggin 2013). Honey bees forage ≤5 km from their hive and can therefore pollinate a diversity of garden plants across the urban landscape. Green roofs may also provide habitat for native bee species that are themselves important pollinators, despite being more cryptic than the honey bee (Oberndorfer et al. 2007; Colla et al. 2009; Tonietto et al. 2011).

Green walls or green facades use vegetation to cover vertical city surfaces. They are comprised of climbing plants growing either directly against the external walls of buildings or on support structures attached to walls, including trellises, steel cables, wire and mesh. Green facades offer a number of benefits in the urban environment, including shading, cooling (through evapotranspiration), increased reflectance of solar radiation, a reduction in wind speed at the surface of the wall, absorption of atmospheric pollutants and habitat for biodiversity (Currie and Bass 2008; Köhler et al. 2008; Hunter et al. 2014). Green facades that use support structures also trap a layer of air between these structures and the wall, which acts to insulate the wall and its building from temperature extremes in both summer and winter (Hunter et al. 2014).

Green facades can be especially useful in narrow city streets (also known as urban canyons) where there is little space to plant trees at ground level (Norton et al. 2015). They provide maximum thermal benefits when planted on west-facing walls in hot, dry climates (Hunter et al. 2014). However, microclimatic conditions (such as temperature and wind speed) on urban walls can be extreme, so a careful choice of plants, growing medium, and (in some cases) irrigation system is required to ensure that green facades thrive. Plants may also spontaneously colonize human-constructed walls in cities, sometimes forming taxonomically and structurally diverse communities and supporting rare or threatened species (Francis 2011). For example, plant communities dominated by strangler figs have developed on stone retaining walls in Hong Kong, providing greenery in an otherwise stark urban landscape (Jim 2014). Seeds brought by frugivorous birds and bats germinate in joints between masonry blocks in the retaining walls. As the fig trees grow, they grip the walls (which act as surrogate host trees).

7.3 Novel habitats, novel ecosystems

Urbanization creates a range of novel habitats for animals and plants, including features such as road corridors, railways, buildings, bridges, recreational parks, private gardens, constructed wetlands, garden ponds, artificial sea walls, and other marine structures such as piers and pontoons. These habitats provide important resources for biodiversity – soil, nutrients, food, water, shelter, sites for nesting, roosting and breeding, or substrates on which to settle. For certain species that can exploit these novel habitats, they may partially compensate for the loss of natural habitats following urbanization. For example, many marine plants and animals can successfully colonize sea walls, including algae, sessile invertebrates such as tubeworms, barnacles and mussels, and mobile invertebrates such as limpets, crabs and snails (Chapman and Blockley 2009; Ng et al. 2012). While a study in Sydney Harbour found vertical, intertidal sea walls to support fewer species in general (and fewer mobile invertebrates in particular) than natural, gently-sloping intertidal shores, the sea walls still provided habitat for >100 species (Chapman 2003). Marine structures in urban areas can also act as habitat for fish, and in certain circumstances may function as artificial reefs. Pillars supporting two oil jetties in the Red Sea at Eilat, Israel attracted 146 species of fish belonging to 35 families, including resident, territorial species and visitors (Rilov and Benayahu 1998). This diversity compared favourably with that of degraded coral reefs nearby. The jetty pillars were made of steel and partially surrounded by barbed wire; the greater the coverage of wire on a pillar, the greater the abundance and diversity of fish. The space between the barbed wire and the pillar provided shelter for juvenile fish, reducing predation risk (Rilov and Benayahu 1998).

A variety of native and introduced reptile species in the West Indies dwell in and on buildings (also known as edificarian habitats) or piles of municipal waste that offer shelter and an abundance of arthropod prey (Powell and Henderson 2008). Private gardens (patios) in León, Nicaragua provide important habitat for the black spiny-tailed iguana *Ctenosaura similis* (Gonzalez-Garcia et al. 2009; Figure 7.6). Known as the fastest lizard on earth, this iguana can run at speeds up to 34.6 $km\,h^{-1}$ (Garland 1984), and males grow up to 1.5 m long. Built in an area that historically supported tropical dry forest, León has a higher tree cover than the surrounding agricultural land, and patios comprise more than 85% of the total green space in the city. Large patios with tall trees, food plants favoured by the iguanas and permeable fences were found to support the highest abundance

Figure 7.6 Black spiny-tailed iguanas *Ctenosaura similis* can be found in private gardens in León, Nicaragua. Picture has been cropped and converted to black and white. Photograph by Christian Mehlführer. https://commons.wikimedia.org/wiki/File%3ACtenosauraSimilis.jpg. Used under CC BY 2.5 http://creativecommons.org/licenses/by/2.5.

of *C. similis* (Gonzalez-Garcia et al. 2009). Adult iguanas spend much of their time high up in trees or on the roofs of dwellings, perhaps avoiding humans who hunt them for food. One quarter of the patios surveyed contained iguana burrows, half of which were built in piles of rubbish and/or construction materials. These few examples show how novel habitats can be exploited by urban-dwelling species in marine and terrestrial environments.

Novel ecosystems (sometimes known as emerging ecosystems) are characterized by the presence of new combinations of species, new relative abundances of species, and possible changes in ecosystem function as a result of human actions (Chapin and Starfield 1997; Milton 2003; Hobbs et al. 2006). They are considered to be novel because they have no historical analogues (e.g., Seastedt et al. 2008; Kowarik 2011; Hobbs et al. 2013; Graham et al. 2014). Novel ecosystems – including novel urban ecosystems – may be difficult or impossible to restore to their previous state, leading to a call for correspondingly novel management approaches to maximize their biodiversity and the ecosystem services they provide rather than expending resources trying to recreate the ecosystems of the past (Hobbs et al. 2006, 2009, 2013; Seastedt et al. 2008). The concept of novel ecosystems has generated a measure of controversy in the field of restoration ecology, with some authors arguing that it is poorly defined, confusing and potentially damaging for conservation (Aronson et al. 2014b;

Murcia et al. 2014). Concern has been raised that the classification of certain ecosystems as novel ecosystems may lead to the abandonment of efforts to restore them when important elements of them are, in fact, salvageable (Aronson et al. 2014b; Murcia et al. 2014). However, the idea that novel urban ecosystems have present ecological value despite their altered state (including the presence of non-native species) and that with suitable management they have the potential to provide further ecological value is a powerful one (Standish et al. 2013). This value may be in the form of habitat for biodiversity, provision of ecosystem services, or places and spaces for city dwellers to interact with nature.

7.4 Summary

The conservation of biodiversity and provision of ecosystem services in cities are not without their challenges, but the benefits of these dual endeavours for humans and other species that live in cities are potentially enormous. Valuable strategies for enhancing biodiversity and ecosystem services in cities include the integration of urban ecology with urban planning and design; protection of important, biodiverse habitats and landscape features; an emphasis on increasing the total cover of trees and other vegetation in cities, as well as increasing the equity of their distribution across cities; the maintenance or improvement of landscape connectivity; and the reclamation of small urban spaces as places for nature. Embracing the present value and future potential of urban ecosystems, despite their altered state, will enable us to conserve a broader diversity of species and reap the benefits of ecosystem services, including shading, cooling, amelioration of atmospheric pollution, nutrient and hydrological cycling and food production. As the world's climate changes, these services will not just improve the amenity and liveability of our cities, but may mean the difference between life and death for many urban dwellers.

Study questions

1. Why are urban areas important for the conservation of biological diversity?
2. Consider a threatened species or ecological community that is still persisting in an urban or peri-urban area near you. Is it vulnerable to further urban expansion? What steps could be taken to protect it?
3. Ecosystem services are essential for the liveability of cities – discuss.
4. How can urban ecology be integrated with urban planning and design to improve biodiversity conservation and the provision of ecosystem services in cities?

5 Explain the importance of the following in urban environments: biodiverse landscape features, transpiring vegetation, and landscape connectivity.
6 What are urban adapters, avoiders and exploiters? How would you recognize each group?
7 Urban ecosystems are novel ecosystems – discuss.

References

Ahern, J. (2013) Urban landscape sustainability and resilience: the promise and challenges of integrating ecology with urban planning and design. *Landscape Ecology*, **28**, 1203–1212.

Akinnifesi, F.K., Sileshi, G.W., Ajayi, O.C. *et al.* (2010) Biodiversity of the urban homegardens of São Luís city, Northeastern Brazil. *Urban Ecosystems*, **13**, 129–146.

Alvey, A.A. (2006) Promoting and preserving biodiversity in the urban forest. *Urban Forestry and Urban Greening*, **5**, 195–201.

Armstrong, A.J. (2008) Translocation of black-headed dwarf chameleons *Bradypodion melanocephalum* in Durban, KwaZulu-Natal, South Africa. *African Journal of Herpetology*, **57**, 29–41.

Aronson, M.F.J., La Sorte, F.A., Nilon, C.H. *et al.* (2014a) A global analysis of the impacts of urbanization on bird and plant diversity reveals key anthropogenic drivers. *Proceedings of the Royal Society B*, **281**, 20133330. http://dx.doi.org/10.1098/rspb.2013.3330

Aronson, J., Murcia, C., Katten, G.H. *et al.* (2014b) The road to confusion is paved with novel ecosystem labels: A reply to Hobbs et al. *Trends in Ecology and Evolution*, **29**, 646–647.

Balmford, A., Bruner, A., Cooper, P. *et al.* (2002) Economic reasons for conserving wild nature. *Science*, **297**, 950–953.

Barthel, S., Colding, J., Elmqvist, T. *et al.* (2005) History and local management of a biodiversity-rich, urban cultural landscape. *Ecology and Society* **10**, 10. www.ecologyandsociety.org/vol10/iss2/art10/

Beier, P. (1995) Dispersal of juvenile cougars in fragmented habitat. *The Journal of Wildlife Management*, **59**, 228–237.

Beier, P. and Noss, R.F. (1998) Do habitat corridors provide connectivity? *Conservation Biology*, **12**, 1241–1252.

Bekessy, S.A. and Gordon, A. (2007) Nurturing nature in the city, in *Steering Sustainability in an Urbanizing World: Policy* (ed A. Nelson), Practice and Performance. Ashgate Publishing, Hampshire, pp. 227–238.

Bekessy, S.A., White, M., Gordon, A. *et al.* (2012) Transparent planning for biodiversity and development in the urban fringe. *Landscape and Urban Planning*, **108**, 140–149.

Benedict, M.A. and McMahon, E.T. (2006) *Green Infrastructure: Smart Conservation for the 21st Century*, The Sprawl Watch Clearinghouse, Washington, DC.

Benítez, G., Perez-Vazquez, A., Nava-Tablada, M. *et al.* (2012) Urban expansion and the environmental effects of informal settlements on the outskirts of Xalapa city, Veracruz, Mexico. *Environment and Urbanization*, **24**, 149–166.

Bjurlin, C.D., Cypher, B.L., Wingert, C.M. *et al.* (2005) *Urban Roads and the Endangered San Joaquin Kit Fox*. FHWA/CA/IR-2006/01. California Department of Transportation,

Fresno. http://www.dot.ca.gov/newtech/researchreports/2002-2006/2005/urban_roads_and_the_endangered_san_joaquin_kit_fox.pdf (Accessed 15/02/2015)

Bolund, P. and Hunhammer, S. (1999) Ecosystem services in urban areas. *Ecological Economics*, **29**, 293–301.

Bonier, F., Martin, P.R., and Wingfield, J.C. (2007) Urban birds have broader environmental tolerance. *Biology Letters*, **3**, 670–673.

Braaker, S., Ghazoul, J., Obrist, M.K. *et al.* (2014) Habitat connectivity shapes urban arthropod communities: the key role of green roofs. *Ecology*, **95**, 1010–1021.

Breed, C., Cilliers, S., and Fisher, R. (2014) Role of landscape designers in promoting a balanced approach to green infrastructure. *Journal of Urban Planning and Development*, **141**, doi: 10.1061/(ASCE)UP.1943-5444.0000248

Brunner, J. and Cozens, P. (2012) Where have all the trees gone? Urban consolidation and the demise of urban vegetation: a case study from Western Australia. *Planning, Practice and Research*, **28**, 231–255.

Calhoun, A.J.K., Arrigoni, J., Brooks, R.P. *et al.* (2014) Creating successful vernal pools: a literature review and advice for practitioners. *Wetlands*, **34**, 1027–1038.

Cameron, R.W.F., Blanusa, T., Taylor, J.E. *et al.* (2012) The domestic garden – its contribution to urban green infrastructure. *Urban Forestry and Urban Greening*, **11**, 129–137.

Carpaneto, G.M., Mazziotta, A., Coletti, G. *et al.* (2010) Conflict between insect conservation and public safety: the case study of a saproxylic beetle (*Osmoderma eremita*) in urban parks. *Journal of Insect Conservation*, **14**, 555–565.

Carrus, G., Scopelliti, M., Lafortezza, R. *et al.* (2015) Go greener, feel better? The positive effects of biodiversity on the well-being of individuals visiting urban and peri-urban green areas. *Landscape and Urban Planning*, **134**, 221–228.

Castleton, H.F., Stovin, V., Beck, S.B.M. *et al.* (2010) Green roofs; building energy savings and the potential for retrofit. *Energy and Buildings*, **42**, 1582–1591.

Chaker, A., El-Fadl, K., Chamas, L. *et al.* (2006) A review of strategic environmental assessment in 12 selected counties. *Environmental Impacts Assessment Review*, **26**, 15–56.

Chapin, F.S. and Starfield, A.M. (1997) Time lags and novel ecosystems in response to transient climatic change in arctic Alaska. *Climate Change*, **35**, 449–461.

Chapman, M.G. (2003) Paucity of mobile species on constructed seawalls: effects of urbanization on biodiversity. *Marine Ecology Progress Series*, **264**, 21–29.

Chapman, M.G. and Blockley, D.J. (2009) Engineering novel habitats on urban infrastructure to increase intertidal biodiversity. *Oecologia*, **161**, 625–235.

Cilliers, S. and Siebert, S.J.. (2012) Urban ecology in Cape Town: South African comparisons and reflections. *Ecology and Society* **17**, 33. http://www.ecologyandsociety.org/vol17/iss3/art33/

Cilliers, S., du Toit, M., Cilliers, J. *et al.* (2014) Sustainable urban landscapes: South African perspectives on transdisciplinary possibilities. *Landscape and Urban Planning*, **125**, 260–270.

Clarke, L.W., Jenerette, G.D., and Davila, A. (2013) The luxury of vegetation and the legacy of tree biodiversity in Los Angeles, CA. *Landscape and Urban Planning*, **116**, 48–59.

Clarke, L.W., Li, L., Jenerette, G.D. *et al.* (2014) Drivers of plant biodiversity and ecosystem service production in home gardens across the Beijing Municipality of China. *Urban Ecosystems*, **17**, 741–760.

Coates, G. (2013) Sustainable urbanism: Creating resilient communities in the age of peak oil and climate destabilization, in *Environmental Policy is Social Policy – Social Policy is Environmental Policy* (ed I. Wallimann), Springer, New York, pp. 81–101.

Colding, J. (2007) 'Ecological land-use complementation' for building resilience in urban ecosystems. *Landscape and Urban Planning*, **81**, 46–55.

Colding, J. and Folke, C. (2009) The role of golf courses in biodiversity conservation and ecosystem management. *Ecosystems*, **12**, 191–206.

Colla, S.R., Willis, E., and Packer, L. (2009) Can green roofs provide habitat for urban bees (*Hymenoptera: Apidae*)? *Cities and the Environment*, **2**, 1–12.

Corbett, J. and Corbett, M. (2000) *Designing Sustainable Communities: Learning From Village Homes*, Island Press, Washington DC.

Costanza, R., de Groot, R., Sutton, P. *et al.* (2014) Changes in the global value of ecosystem services. *Global Environmental Change*, **26**, 152–158.

Coutts, A.M., Tapper, N.J., Beringer, J. *et al.* (2013) Watering our cities. The capacity for Water Sensitive Urban Design to support urban cooling and improve human thermal comfort in the Australian context. *Progress in Physical Geography*, **37**, 2–28.

Croci, S., Butet, A., and Clergeau, P. (2008) Does urbanization filter birds on the basis of their biological traits? *The Condor*, **110**, 223–240.

Crook, C. and Clapp, R.A. (1998) Is market-oriented forest conservation a contradiction in terms? *Environmental Conservation*, **25**, 131–145.

Currie, B.A. and Bass, B. (2008) Estimates of air pollution mitigation with green plants and green roofs using the UFORE model. *Urban Ecosystems*, **11**, 409–422.

Dearborn, D.C. and Kark, S. (2010) Motivations for conserving urban biodiversity. *Conservation Biology*, **24**, 432–440.

de Groot, R., Brander, L., van der Ploeg, S. *et al.* (2012) Global estimates of the value of ecosystems and their services in monetary units. *Ecosystem Services*, **1**, 50–61.

DEPI (2013a) *Sub-regional Species Strategy for the Growling Grass Frog*. Victorian Government Department of Environment and Primary Industries, Melbourne. http://www.depi.vic.gov.au/__data/assets/pdf_file/0016/204343/GGF-SSS-text-only.pdf (accessed 25/06/2015)

DEPI (2013b) *Biodiversity Conservation Strategy for Melbourne's Growth Corridors*. Victorian Government Department of Environment and Primary Industries, Melbourne. http://www.depi.vic.gov.au/__data/assets/pdf_file/0012/204330/BCS-no-maps-PtA.pdf (accessed 25/06/2015)

Diaz-Porras, D.F., Gaston, K.J., and Evans, K.L. (2014) 110 years of change in urban tree stocks and associated carbon storage. *Ecology and Evolution*, **4**, 1413–1422.

DSE (2009) *Delivering Melbourne's Newest Sustainable Communities: Strategic Impact Assessment Report for Environment Protection and Biodiversity Conservation Act 1999*, Victorian Government Department of Sustainability and Environment, Melbourne.

DSE (2011) *Western Grassland Reserves: Grassland Management Targets and Adaptive Management (2011)*. Victorian Government Department of Sustainability and Environment, Melbourne. http://www.depi.vic.gov.au/__data/assets/pdf_file/0017/204371/WGR_TAM.pdf (accessed 25/06/2015)

Dwivedi, P., Rathore, C.S., and Dubey, Y. (2009) Ecological benefits of urban forestry: The case of Kerwa Forest Area (KFA), Bhopal, India. *Applied Geography*, **29**, 194–200.

Eppink, F.V., van den Bergh, J.C.J.M., and Rietveld, P. (2004) Modelling biodiversity and land use: urban growth, agriculture and nature in a wetland area. *Ecological Economics*, **51**, 201–216.

Escobedo, F.J., Nowak, D.J., Wagner, J.E. *et al.* (2006) The socioeconomics and management of Santiago de Chile's public urban forests. *Urban Forestry and Urban Greening*, **4**, 105–114.

Fischer, J. and Lindenmayer, D.B. (2000) An assessment of the published results of animal relocations. *Biological Conservation*, **96**, 1–11.

Francis, R.A. (2011) Wall ecology: A frontier of urban biodiversity and ecological engineering. *Progress in Physical Geography*, **35**, 43–63.

Fu, W., Liu, S., Degloria, S.D. *et al.* (2010) Characterizing the "fragmentation-barrier" effect of road networks on landscape connectivity: a case study in Xishuangbanna, Southwest China. *Landscape and Urban Planning*, **95**, 122–129.

Fuller, R.A., Irvine, K.N., Devine-Wright, P. *et al.* (2007) Psychological benefits of greenspace increase with biodiversity. *Biology Letters*, **3**, 390–394.

Gago, E.J., Roldan, J., Pacheco-Torres, R. *et al.* (2013) The city and urban heat islands: A review of strategies to mitigate adverse effects. *Renewable and Sustainable Energy Reviews*, **25**, 749–758.

Garland, T. (1984) Physiological correlates of locomotory performance in a lizard: an allometric approach. *American Journal of Physiology*, **247**, 806–815.

Garrard, G. and Bekessy, S. (2014) Land use and land management, in *Australian Environmental Planning: Challenges and Future Prospects* (eds J. Byrne, N. Sipe, and J. Dodson), Routledge, Abingdon, pp. 61–72.

Gaston, K.J., Fuller, R.A., Loram, A. *et al.* (2007) Urban domestic gardens (XI): variation in urban wildlife gardening in the United Kingdom. *Biodiversity and Conservation*, **16**, 3227–3238.

Gaston, K.J., Warren, P.H., Thompson, K. *et al.* (2005) Urban domestic gardens (IV): the extent of the resource and its associated features. *Biological Conservation*, **14**, 3327–3349.

Germano, J.M. and Bishop, P.J. (2009) Suitability of amphibians and reptiles for translocation. *Conservation Biology*, **23**, 7–15.

Gibbs, J.P. and Shriver, W.G. (2002) Estimating the effects of road mortality on turtle populations. *Conservation Biology*, **16**, 1647–1652.

Gibbs, J.P. and Shriver, W.G. (2005) Can road mortality limit populations of pool-breeding amphibians? *Wetland Ecology and Management*, **13**, 281–289.

Goddard, M.A., Dougill, A.J., and Benton, T.G. (2010) Scaling up from gardens: biodiversity conservation in urban environments. *Trends in Ecology and Evolution*, **25**, 90–98.

Goddard, M.A., Dougill, A.J., and Benton, T.G. (2013) Why garden for wildlife? Social and ecological drivers, motivations and barriers for biodiversity management in residential landscape. *Ecological Economics*, **86**, 258–273.

Gonzalez-Garcia, A., Belliure, J., Gomez-Sal, A. *et al.* (2009) The role of urban greenspaces in fauna conservation: the case of the iguana *Ctenosaura similis* in the 'patios' of León city, Nicaragua. *Biodiversity and Conservation*, **18**, 1909–1920.

Gordon, A., Simondson, D., White, M. *et al.* (2009) Integrating conservation planning and landuse planning in urban landscapes. *Landscape and Urban Planning*, **91**, 183–194.

Graham, N.A.J., Cinner, J.E., Norstrom, A.V. *et al.* (2014) Coral reefs as novel ecosystems: embracing new futures. *Current Opinion in Environmental Sustainability*, **7**, 9–14.

Grant, B.W., Middendorf, G., Colgan, M.J. *et al.* (2011) Ecology of urban amphibians and reptiles: urbanophiles, urbanophobes, and the urbanoblivious, in *Urban Ecology: Patterns, Processes, and Applications* (ed J. Niemelä), Oxford University Press, Oxford, pp. 167–178.

Grewal, S.S. and Grewal, P.S. (2012) Can cities become self-reliant in food? *Cities*, **29**, 1–11.

Gupta, R.B., Chaudhari, P.R., and Wate, S.R. (2008) Floristic diversity in urban forest area of NEERI Campus, Nagpur, Maharashtra (India). *Journal of Environmental Science & Engineering*, **50**, 55–62.

Hahs, A.K., McDonnell, M.J., McCarthy, M.A. *et al.* (2009) A global synthesis of plant extinction rates in urban areas. *Ecology Letters*, **12**, 1165–1173.

Hale, J.M., Heard, G.W., Smith, K.L. *et al.* (2013) Structure and fragmentation of growling grass frog metapopulations. *Conservation Genetics*, **14**, 313–322.

Hall, T. (2011) What has happened to the great Aussie backyard? *The Conversation*. http://theconversation.com/what-has-happened-to-the-great-aussie-backyard-4506 (accessed 06/06/2015)

Hamer, A.J., van der Ree, R., Mahony, M.J. *et al.* (2014) Usage rates of an under-road tunnel by three Australian frog species: implications for road mitigation. *Animal Conservation*, **17**, 379–387.

Harlan, S.L., Brazel, A.J., Jenerette, G.D. *et al.* (2007) In the shade of affluence: the inequitable distribution of the urban heat island. *Research in Social Problems and Public Policy*, **15**, 173–202.

Heard, G.W., McCarthy, M.A., Scroggie, M.P. *et al.* (2013) A Bayesian model of metapopulation viability, with application to an endangered amphibian. *Diversity and Distributions*, **19**, 555–566.

Heard, G.W., Scroggie, M.P., and Malone, B.S. (2012) The life history and decline of the threatened Australian frog, *Litoria raniformis*. *Austral Ecology*, **37**, 276–284.

Hes, D. and Plessis, C.D. (2015) *Designing for Hope: Pathways to Regenerative Sustainability*, Routledge, Abingdon.

Hetrick, S., Chowdhury, R.R., Brondizio, E. *et al.* (2013) Spatiotemporal patterns and socioeconomic contexts of vegetative cover in Altamira City, Brazil. *Land*, **2**, 774–796.

Hilty, J.A., Lidlicker, W.Z., and Merenlender, A.M. (2006) *Corridor Ecology: The Science and Practice of Linking Landscapes for Biodiversity Conservation*, Island Press, Washington DC.

Hobbs, R.J., Higgs, E., and Hall, C. (2013) *Novel Ecosystems: Intervening in the New Ecological World Order*, Wiley-Blackwell, Oxford.

Hobbs, R.J., Higgs, E., and Harris, J.A. (2009) Novel ecosystems: implications for conservation and restoration. *Trends in Ecological and Evolution*, **24**, 599–605.

Hobbs, R.J., Arico, S., Aronson, J. *et al.* (2006) Novel ecosystems: theoretical and management aspects of the new ecological world order. *Global Ecology and Biogeography*, **15**, 1–7.

Hunter, A.M., Williams, N.S.G., Rayner, J.P. *et al.* (2014) Quantifying the thermal performance of green facades: A critical review. *Ecological Engineering*, **63**, 102–113.

Hunter, M.R. and Hunter, M.D. (2008) Designing for conservation of insects in the built environment. *Insect Conservation and Diversity*, **1**, 189–196.

IAIA (International Association for Impact Assessment). (2002) Strategic environmental assessment performance criteria. *IAIA Special Publication Series*, 1. http://www.iaia.org/publicdocuments/special-publications/sp1.pdf?AspxAutoDetectCookieSupport=1 (accessed 25/06/2015)

Indykiewicz, P., Jerzak, L., Bohner, J. *et al.* (2011) Urban Fauna, in *Studies of Animal Biology, Ecology and Conservation in European Cities*. University of Technology and Life Sciences in Bydgoszcz, Bydgoszcz.

Ives, C.D., Lentini, P.E., Threlfall, C.G. *et al.* (2016) Cities are hotspots for threatened species. *Global Ecology and Biogeography* **25**, 117–126.

Jacobs, B., Mikhailovich, N. and Delaney, C. (2014) *Benchmarking Australia's Urban Tree Canopy: An i-Tree Assessment*. Institute for Sustainable Futures, University of Technology Sydney. http://202020vision.com.au/media/7141/benchmarking_australias_urban_tree_canopy.pdf (accessed 25/06/2015)

Jaganmohan, M., Vailshery, L.S., Gopal, D. *et al.* (2012) Plant diversity and distribution in urban domestic gardens and apartments in Bangalore, India. *Urban Ecosystems*, **15**, 911–925.

James, C. (2009) Australian homes are the biggest in the world. *ComSec Economic Insights*, November 30, 2009. https://img.yumpu.com/37485169/1/358x507/australian-homes-are-biggest-in-the-world-comsec.jpg (accessed 12/11/2015)

Jim, C.Y. (2014) Ecology and conservation of strangler figs in urban wall habitats. *Urban Ecosystems*, **17**, 405–426.

Jim, C.Y. and Chen, S.S. (2003) Comprehensive greenspace planning based on landscape ecology principles in compact Nanjing city, China. *Landscape and Urban Planning*, **65**, 95–116.

Jo, H. (2002) Impacts of urban greenspace on offsetting carbon emissions for middle Korea. *Journal of Environmental Management*, **64**, 115–126.

Kadas, G. (2002) Rare invertebrates colonizing green roofs in London. *Urban Habitats*, **4**, 66–86.

Karvonen, A. (2011) *Politics of Urban Runoff: Nature, Technology, and the Sustainable City*, MIT Press, Cambridge.

Kendal, D., Williams, N.S.G., and Williams, K.J.H. (2010) Harnessing diversity in gardens through individual decision makers. *Trends in Ecology and Evolution*, **25**, 201–202.

Kendal, D., Williams, N.S.G., and Williams, K.J.H. (2012) Drivers of diversity and tree cover in gardens, parks, and streetscapes in an Australian city. *Urban Forestry and Urban Greening*, **11**, 257–265.

Koehler, S.L. and Gilmore, D.C. (2015) Translocation of the threatened Growling Grass Frog *Litoria raniformis*: a case study. *The Australian Zoologist*, **37**, 321–336.

Köhler, M. (2008) Green facades – a view back and some visions. *Urban Ecosystems*, **11**, 423–436.

Kowarik, I. (2011) Novel urban ecosystems, biodiversity, and conservation. *Environmental Pollution*, **159**, 1974–1983.

Knapp, S., Kuhn, I., Schweiger, O. *et al.* (2008) Challenging urban species diversity: contrasting phylogenetic patterns across plant functional groups in Germany. *Ecology Letters*, **11**, 1054–1064.

Le Roux, D.S., Ikin, K., Lindenmayer, D.B. *et al.* (2014) The future of large old trees in urban landscapes. *PLoS ONE* **9**, e99403.

Lee, K.E., Williams, K.J.H., Sargent, L.D. *et al.* (2015) 40-second green roof views sustain attention: the role of micro-breaks in attention restoration. *Journal of Environmental Psychology*, **42**, 182–189.

Lehvävirta, S. (2007) Non-anthropogenic dynamic factors and regeneration of (hemi)boreal urban woodlands – synthesising urban and rural ecological knowledge. *Urban Forestry and Urban Greening*, **6**, 119–134.

Lenda, M., Skorka, P., Moron, D. *et al.* (2012) The importance of the gravel excavation industry for the conservation of grassland butterflies. *Biological Conservation*, **148**, 180–190.

Levick, L.R., D.C. Goodrich, M. Hernandez et al. 2008. *The Ecological and Hydrological Significance of Ephemeral and Intermittent Streams in the Arid and Semi-Arid American Southwest.* US Environmental Protection Agency, Office of Research and Development. http://www.epa.gov/esd/land-sci/pdf/EPHEMERAL_STREAMS_REPORT_Final_508-Kepner.pdf (accessed 25/06/2015)

Li, F., Wang, R., Paulussen, J. *et al.* (2005) Comprehensive concept planning of urban greening based on ecological principles: a case study in Beijing, China. *Landscape and Urban Planning*, **72**, 325–336.

Lindenmayer, D.B., Laurance, W.F., and Franklin, J.F. (2012) Global decline in large old trees. *Science*, **338**, 1305–1306.

Livesley, S.J., Dougherty, B.J., Smith, A.J. *et al.* (2010) Soil-atmosphere exchange of carbon dioxide, methane, and nitrous oxide in urban garden systems: impact of irrigation, fertiliser, and mulch. *Urban Ecosystems*, **13**, 273–293.

Loram, A., Warren, P.H., and Gaston, K.J. (2008) Urban domestic gardens (XIV): the characteristics of gardens in five cities. *Environmental Management*, **42**, 361–376.

Mahony, M. (1999) Review of the declines and disappearances within the bell frog species group (*Litoria aurea* species group) in Australia, in *Declines and Disappearances of Australian Frogs* (ed A. Campbell), Environment Australia, Canberra, pp. 81–93.

Marchand, M.N. and Litvaitis, J.A. (2004) Effects of habitat features and landscape composition on the population structure of a common aquatic turtle in a region undergoing rapid development. *Conservation Biology*, **18**, 758–767.

Maric, T., Zaninovic, J., and Scitaroci, B.B.O. (2008) Landscape as a connection – beyond boundaries. in M. Schrenk, V.V. Popovich, P. Zeile and P. Elisei (eds), *REAL CORP 2013: Planning Times*. Proceedings of the 18th International Conference on Urban Planning, Regional Development and Information Society. CORP – Competence Center of Urban and Regional Planning, Schwechat, Austria, pp. 497–506 http://corp.at/fileadmin/proceedings/CORP2013_proceedings.pdf (accessed 25/06/2015).

Marriage, G. (2010) Minimum vs maximum: size and the New Zealand house. *2010 Australasian Housing Researchers' Conference*. http://www.academia.edu/3712795/Minimum_vs_Maximum_size_and_the_New_Zealand_House_-_first_published_in_the_Australasian_Housing_Researchers_Conference_Auckland_2010 (accessed 25/06/2015)

Martell, M.S., Englund, J.V., and Tordoff, H.B. (2002) An urban osprey population established by translocation. *Journal of Raptor Research*, **36**, 91–96.

McDonald, R.I., Kareiva, P., and Forman, R.T.T. (2008) The implication of current and future urbanization for global protected areas and biodiversity conservation. *Biological Conservation*, **141**, 1695–1703.

McKinney, M.L. (2002) Urbanization, biodiversity, and conservation. *BioScience*, **52**, 883–890.

McPherson, E.G. and Rowntree, R.A. (1993) Energy conservation potential of urban tree planting. *Journal of Arboriculture*, **19**, 321–331.

McPherson, E.G., Nowak, D., Heisler, G. *et al.* (1997) Quantifying urban forest structure, function, and value: the Chicago Urban Forest Climate Project. *Urban Ecosystems*, **1**, 49–61.

MEA (Millennium Ecosystem Assessment) (2005) *Ecosystems and Human Well-Being*, World Resources Institute, Washington DC.

Mentens, J., Raes, D., and Hermy, M. (2006) Green roofs as a tool for solving the rainwater runoff problem in the urbanized 21st century? *Landscape and Urban Planning*, **77**, 217–226.

Milton, S.J. (2003) Emerging ecosystems: a washing-stone for ecologists, economists and sociologists? *South African Journal of Science*, **99**, 404–406.

Mitchell, J.C., Jung Brown, R.E., and Bartholomew, B. (eds). (2008) *Urban Herpetology*. Herpetological Conservation 3. Society for the Study of Amphibians and Reptiles, Salt Lake City.

Murcia, C., Aronson, J., Kattan, G.H. *et al.* (2014) A critique of the 'novel ecosystem' concept. *Trends in Ecology and Evolution*, **29**, 548–553.

Newbound, M., Bennett, L.T., Tibbits, J. *et al.* (2012) Soil chemical properties, rather than landscape context, influence woodland fungal communities along an urban-rural gradient. *Austral Ecology*, **37**, 236–247.

Ng, C.S.L., Chen, D., and Chou, L.M. (2012) Hard coral assemblages on seawalls in Singapore. *Contributions to Marine Science*, **2012**, 75–79.

Ng, S.J., Dole, J.W., Sauvajot, R.M. *et al.* (2004) Use of highway undercrossings by wildlife in southern California. *Biological Conservation*, **115**, 499–507.

Norton, B.A., Bosomworth, K., Coutts, A. *et al.* (2013) *Planning for a Cooler Future: Green Infrastructure to Reduce Urban Heat*. Victorian Centre for Climate Change Adaptation Research,

Melbourne. http://www.vcccar.org.au/sites/default/files/publications/VCCCAR%20Green%20Infrastructure%20Guide%20Final.pdf (accessed 25/06/2015)

Norton, B.A., Coutts, A.M., Livesley, S.J. *et al.* (2015) Planning for cooler cities: A framework to prioritise green infrastructure to mitigate high temperatures in urban landscapes. *Landscape and Urban Planning*, **134**, 127–138.

NSW Department of Planning. (2010) *Sydney Growth Centres Strategic Assessment Program Report*. Department of Planning New South Wales, Sydney. http://www.environment.gov.au/epbc/notices/assessments/pubs/sydney-growth-centres-program-report.pdf (accessed 25/06/2015)

Oberndorfer, E., Lundholm, J., Bass, B. *et al.* (2007) Green roofs as urban ecosystems: ecological structures, functions, and services. *BioScience*, **57**, 823–833.

Ong, B.L. (2003) Green plot ratio: an ecological measure for architecture and urban planning. *Landscape and Urban Planning*, **63**, 197–211.

Parris, K.M. (2006) Urban amphibian assemblages as metacommunities. *Journal of Animal Ecology*, **75**, 757–764.

Parris K.M. (2009) What are strategic impact assessments? *Decision Point* **32**, 4-6. http://decision-point.com.au/wp-content/uploads/2014/12/DPoint_32.pdf (accessed 25/06/2015)

Pauchard, A., Aguayo, M., Peña, E. *et al.* (2006) Multiple effects of urbanization on the biodiversity of developing countries: The case of a fast-growing metropolitan area (Concepción, Chile). *Biological Conservation*, **127**, 272–281.

Pedlowski, M.A., Da Silva, V.A.C., Adell, J.J.C. *et al.* (2002) Urban forest and environmental inequality in Campos dos Goytacazes, Rio de Janeiro, Brazil. *Urban Ecosystems*, **6**, 9–20.

Pietsch, R.S. (1994) The fate of urban Common Brushtail Possums translocated to sclerophyll forest, in *Reintroduction Biology of Australian and New Zealand Fauna* (ed M. Serena), Surrey Beatty and Sons, Chipping Norton, pp. 239–246.

Polasky, S., Nelson, E., Camm, J. *et al.* (2008) Where to put things? Spatial land management to sustain biodiversity and economic returns. *Biological Conservation*, **141**, 1505–1524.

Polk, M.H., Young, K.R., and Crews-Meyer, K.A. (2005) Biodiversity conservation implications of landscape change in an urbanizing desert of Southwestern Peru. *Urban Ecosystems*, **8**, 313–334.

Powell, P., and Henderson, R.W. (2008) Urban herpetology in the West Indies. in J.C. Mitchell, R.E. Jung Brown, and B. Bartholomew (eds), *Urban Herpetology*. Herpetological Conservation **3**. Society for the Study of Amphibians and Reptiles, Salt Lake City, pp. 389–404.

Pyke, G.H. (2002) *A review of the biology of the southern bell frog Litoria raniformis (Anura: Hylidae) Australian Zoologist*, **32**, 32–48.

Rilov, G. and Benayahu, Y. (1998) Vertical artificial structures as an alternative habitat for coral reef fishes in disturbed environments. *Marine Environmental Research*, **45**, 431–451.

Rosenzweig, M.L. (2003a) *Win-Win Ecology: How the Earth's Species Can Survive in the Midst of Human Enterprise*, Oxford University Press, Oxford.

Rosenzweig, M.L. (2003b) Reconciliation ecology and the future of species diversity. *Oryx*, **37**, 194–205.

Rudd, H., Vala, J., and Schaefer, V. (2002) Importance of backyard habitat in a comprehensive biodiversity conservation strategy: a connectivity analysis of urban green spaces. *Restoration Ecology*, **10**, 368–375.

Rumming, K. (ed). (2004) *Hannover Kronsberg Handbook, Planning and Realisation*. City of Hannover, Germany.

Saggin, G. (2013) Urban beehives on the increase as global bee numbers decline. http://www.abc.net.au/news/2013-11-15/urban-beehive-movement-in-australia-and-around-the-world/5093764 (accessed 25/06/2015)

Savard, J.L., Clergeau, P., and Mennechez, G. (2000) Biodiversity concepts and urban ecosystems. *Landscape and Urban Planning*, **48**, 131–142.

Seastedt, T.R., Hobbs, R.J., and Suding, K.N. (2008) Management of novel ecosystems: are novel approaches required? *Frontiers in Ecology and the Environment*, **6**, 547–553.

Secretariat of the Convention on Biological Diversity. (2012) *Cities and Biodiversity Outlook*. Secretariat of the Convention on Biological Diversity, Montreal. http://www.cbd.int/doc/health/cbo-action-policy-en.pdf (accessed 25/06/2015)

Seto, K.C., Guneralp, B., and Hutyra, L.R. (2012) Global forecasts of urban expansion to 2030 and direct impacts on biodiversity and carbon pools. *Proceedings of the National Academy of Sciences of the United States of America*, **109**, 16083–16088.

Shanahan, D.F., Lin, B.B., Gaston, K.J. *et al.* (2014) Socio-economic inequality in access to nature on public and private lands: A case study from Brisbane, Australia. *Landscape and Urban Planning*, **130**, 14–23.

Shariful Islam, K.M. (2004) Rooftop gardening as a strategy of urban agriculture for food security: the case of Dhaka City, Bangladesh. *Acta Horticulturae*, **643**, 241–247.

Snyder, S.A., Miller, J.R., Skibbe, A.M. *et al.* (2007) Habitat acquisition strategies for grassland birds in an urbanizing landscape. *Environmental Management*, **40**, 981–992.

Stagoll, K., Lindenmayer, D.B., Knight, E. *et al.* (2012) Large trees are keystone structures in urban parks. *Conservation Letters*, **5**, 115–122.

Standish, R.J., Hobbs, R.J., and Miller, J.R. (2013) Improving city life: options for ecological restoration in urban landscapes and how these might influence interactions between people and nature. *Landscape Ecology*, **28**, 1213–1221.

Stovin, V.R., Jorgensen, A., and Clayden, A. (2015) Street trees and stormwater management. *Arboriculture Journal: The International Journal of Urban Forestry*, **30**, 297–310.

Sung, C.Y. (2013) Mitigating surface urban heat island by a tree protection policy: A case study of The Woodland, Texas, USA. *Urban Forestry and Urban Greening*, **12**, 474–480.

Tanner, C.J., Adler, F.R., Grimm, N.B. *et al.* (2014) Urban ecology: advancing science and society. *Frontiers in Ecology and the Environment*, **12**, 574–581.

Taylor, L. and Hochuli, D.F. (2015) Creating better cities: how biodiversity and ecosystem functioning enhance urban residents' wellbeing. *Urban Ecosystems*, **18**, 747–762.

Threlfall, C.G., Law, B. and Banks, P.B. (2012) Influence of landscape structure and human modifications on insect biomass and bat foraging activity in an urban landscape. *PLoS ONE* **7**, E38800.

Threlfall, C.G., Williams, N.S.G. , Hahs, A.K. *et al.* (2014) Green havens. *Australian Turfgrass Management* 16.5, September-October 2014. http://www.agcsa.com.au/files/Biodiversity%2016.5%20pg%206-12pdf.pdf (accessed 25/06/2015)

Tonietto, R., Fant, J., Ascher, J. *et al.* (2011) A comparison of bee communities of Chicago green roofs, parks and prairies. *Landscape and Urban Planning*, **103**, 102–108.

Tyrvainen, L., Pauleit, S., Seeland, K. *et al.* (2005) Benefits and uses of urban forests and trees, in *Urban forests and Trees: A Reference Book* (eds C.C. Konijnendijk, K. Nilsson, T.B. Randrup, and J. Schipperijn), Springer, Berlin, pp. 81–114.

Tzoulas, K., Korpela, K., Venn, S. *et al.* (2007) Promoting ecosystem and human health in urban areas using green infrastructure: a literature review. *Landscape and Urban Planning*, **81**, 167–178.

Umwelt. (2013) *Gungahlin Strategic Assessment: Final Assessment Report.* Prepared on behalf of the ACT Economic Development Directorate and ACT Environment and Sustainable Development Directorate, Canberra. http://www.economicdevelopment.act.gov.au/__data/assets/pdf_file/0009/480177/8024_R02_V7-Assessment-Report.pdf (accessed 25/06/2015)

UN-HABITAT. (2010) *The State of African cities: Governance, Inequality and Urban Land Markets.* United Nations Human Settlement Program, Nairobi, Kenya. http://mirror.unhabitat.org/pmss/listItemDetails.aspx?publicationID=3034 (accessed 25/06/2015)

Urbonas, B.R. and Doerfer, J.T. (2005) Stream protection in urban watersheds through master planning. *Journal of Water Science and Technology,* **51**, 239–247.

US Census Bureau. (2014) *Median and Average Square Feet of Floor Area in New Single-Family Houses Completed by Location.* US Census Bureau, Washington DC. https://www.census.gov/construction/chars/pdf/medavgsqft.pdf (accessed 10/07/15)

Uy, P.D. and Nakagoshi, N. (2008) Application of land suitability analysis and landscape ecology to urban greenspace planning in Hanoi, Vietnam. *Urban Forestry and Urban Greening,* **7**, 25–40.

Vallance, S., Perkins, H.C., Bowring, J. *et al.* (2012) Almost invisible: glimpsing the city and its residents in the urban sustainability discourse. *Urban Studies,* **49**, 1695–1710.

van der Grift, E.A., van der Ree, R., Fahrig, L. *et al.* (2013) Evaluating the effectiveness of road mitigation measures. *Biodiversity and Conservation,* **22**, 425–448.

Vergnes, A., Kerbiriou, C., and Clergeau, P. (2013) Ecological corridors also operate in an urban matrix: A test case with garden shrews. *Urban Ecosystems,* **16**, 511–525.

Vergnes, A., Le Viol, I., and Clergeau, P. (2012) Green corridors in urban landscape affect the arthropod communities of domestic gardens. *Biological Conservation,* **154**, 171–178.

von Haaren, C. and Reich, M. (2006) The German way to greenways and habitat networks. *Landscape and Urban Planning,* **76**, 7–22.

Walsh, C.J., Roy, A.H., Feminella, J.W. *et al.* (2005) The urban stream syndrome: current knowledge and the search for a cure. *Journal of North American Benthological Society,* **24**, 706–723.

Wang, Y., Piao, Z.J., Wang, X.Y. *et al.* (2013) Road mortalities of vertebrate species on Ring Changbai Mountain Scenic Highway, Jilin Province, China. *North-Western Journal of Zoology,* **9**, 399–409.

Ward, A.I., Dendy, J., and Cowan, D.P. (2015) Mitigation impacts of roads on wildlife: an agenda for the conservation of priority European protected species in Great Britain. *European Journal of Wildlife Research,* **61**, 199–211.

Weeks, J.R., Hill, A., Stow, D. *et al.* (2007) Can we spot a neighbourhood from the air? Defining neighbourhood structure in Accra, Ghana. *Geojournal,* **69**, 9–22.

White, A.W. and Pyke, G.H. (2008) Frogs on the hop: translocations of Green and Golden Bell Frogs *Litoria aurea* in Greater Sydney. *Australian Zoologist,* **34**, 249–260.

Williams, N.S.G., Morgan, J.W., McCarthy, M.A. *et al.* (2006) Local extinction of grassland plants: the landscape matrix is more important than patch attributes. *Ecology,* **87**, 3000–3006.

Williams, N.S.G., Morgan, J.W., McDonnell, M.J. *et al.* (2005) Plant traits and local extinctions in natural grasslands along an urban-rural gradient. *Journal of Ecology,* **93**, 1203–1213.

Wilson, J.N., Bekessy, S., Parris, K.M. *et al.* (2013) Impacts of climate change and urban development on the spotted marsh frog (*Limnodynastes tasmaniensis*). *Austral Ecology,* **38**, 11–22.

Windmiller, B. and Calhoun, A.J.K. (2008) Conserving vernal pool wildlife in urbanizing landscapes, in *Science and Conservation of Vernal Pools in Northeastern North America* (eds A.J.K. Calhoun and P.G. DeMaynadier), CRC Press, Boca Raton, pp. 235–251.

Wolch, J.R., Byrne, J., and Newell, J.P. (2014) Urban green space, public health, and environmental justice: The challenge of making cities 'just green enough'. *Landscape and Urban Planning*, **125**, 234–244.

Wong, N.H., Tay, S.F., Wong, R. *et al.* (2003) Life cycle cost analysis of rooftop gardens in Singapore. *Building and Environment*, **38**, 499–509.

Yasuda, M. and Koike, F. (2006) Do golf courses provide a refuge for flora and fauna in Japanese urban landscapes? *Landscape and Urban Planning*, **75**, 58–68.

Yencken, D. and Wilkinson, D. (2000) *Resetting the Compass: Australia's Journey Towards Sustainability*, CSIRO Publishing, Melbourne.

CHAPTER 8

Summary and future directions

8.1 Introduction

This book is designed to serve as an introduction to urban ecology. In the preceding chapters, I have provided a synthesis of the ecology of urban environments using examples from the developed and developing world. I have disentangled and examined the biophysical processes that occur during urbanization and explored how these processes work together to influence the characteristics of urban environments (Chapter 2), and the dynamics of populations, communities and ecosystems (Chapters 3–5). I have considered the ecology of human populations in cities and the many ways in which urban environments can affect the health and well-being of their human residents (Chapter 6). I have argued the case for conserving biodiversity and maintaining ecosystem services in urban environments – for the benefit of humans and all the other species that live in cities – and have outlined some practical strategies for achieving these goals (Chapter 7). What remains now is to answer one overarching question and to discuss some future directions for the discipline of urban ecology.

8.2 Do we need a new theory of urban ecology?

The overarching question is this: do we need a new theory of urban ecology, or can we understand the ecology of urban environments using existing ecological theories? One school of thought within urban ecology is that current ecological theories cannot capture the full complexity of urban ecosystems, where natural systems and human systems intersect (sometimes referred to as coupled human-ecological or social-ecological systems; Liu et al. 2007; Alberti 2008; Collins et al. 2011). For example, Alberti (2008) argued that "urban ecosystems exhibit unique properties, patterns and behaviors that arise from a complex coupling of humans and ecological processes", and that existing ecological theories alone are not sufficient to understand their function and dynamics (p. 17). Further, she maintained that we need a theory of urban ecology that builds on a plurality of concepts from multiple disciplines to address the mechanisms underlying the behaviour of urban ecosystems (Alberti 2008). Pickett et al. (2008) stated that the discipline of urban

Ecology of Urban Environments, First Edition. Kirsten M. Parris.

ecology lacks a theory, while Cadenasso and Pickett (2008) temporised slightly, suggesting that urban ecology does not yet have a "complete, mature" theory. Leaving aside the distinction between a complete and incomplete (or immature and mature) theory, these authors raise a number of interesting points regarding the complexity, human domination and uniqueness of urban ecosystems which I explore below.

8.2.1 The complexity of urban ecosystems

It has been argued that urban ecosystems are more complex than other types of ecosystems because of their dynamism, their heterogeneity (across multiple spatial and temporal scales), their nonlinearity, the actions of their diverse human populations, and the existence of social-biophysical feedbacks in the urban environment (Alberti et al. 2003; 2008; Andersson 2006; Pickett et al. 2008; Qureshi et al. 2014). For example, Alberti et al. (2003) suggested that urban researchers need to "address explicitly the complexities of many factors working simultaneously on scales from the individual to the regional and global". However, none of these characteristics is confined to urban ecosystems. Many other ecosystems are dynamic and heterogeneous, with various properties and transitional states that do not follow linear trajectories (e.g., Levin 1998; Pen-Mouratov and Steinberger 2005; Anand et al. 2010). The complexity of many variables operating simultaneously across multiple spatial and temporal scales is nothing out of the ordinary for the discipline of ecology (Levin 1998; Anand et al. 2010); such complexity can be observed in a diversity of ecosystems from deserts, grasslands, woodlands, streams and lakes, to reefs and the open ocean (George et al. 1992; Pen-Mouratov and Steinberger 2005; Hobbs et al. 2007; Boit et al. 2012; Sugihara et al. 2012; Bozec et al. 2013; Laurance et al. 2014).

Every ecosystem on Earth is now influenced in some way by the actions of humans (Vitousek et al. 1997), either directly (e.g., humans gathering plants for food and medicine, hunting wild animals, or changing landscape characteristics through agriculture and production forestry) or indirectly (e.g., anthropogenic climate change and its effects on the polar ice caps and the permafrost in the northern tundra; Wipf et al. 2009; Schuur et al. 2015). The actions of humans in all these ecosystems are influenced by a diversity of complex and dynamic social factors, and each ecosystem responds to the actions of humans in ways that vary over time and space. Given that we understand so much about the motivations and actions of humans, as individuals and collectively in society, one could even argue that urban ecosystems are less complex to study than ecosystems where other species dominate. I suggest that an over-emphasis on the complexity of urban ecosystems (and thus, of urban ecology) may hinder rather than help to extend our understanding of cities, their environments and inhabitants.

8.2.2 The human domination of urban ecosystems

As outlined above, the impact of humans on an ecosystem is a continuous rather than a binary variable, a matter of degree rather than an either/or proposition. Urban ecosystems obviously fall at the right-hand end of this gradient of human impact, but do the dominance of humans and their various actions in cities separate cities from all other ecosystems? As observed by Collins et al. (2000), the activities of humans in cities affect urban ecosystems in many important ways – altering landscapes, mobilizing nutrients, diverting water and energy, changing local climates, and driving certain species extinct while supporting increased populations of others. However, if humans really are a component of nature and the ecosystems they inhabit (McDonnell and Pickett 1997), their actions in shaping and modifying those ecosystems should not be regarded differently to those of non-human species that play an important functional role in shaping other types of ecosystems such as deserts, wetlands or coral reefs. Instead, humans could be viewed as ecosystem engineers in cities (Jones et al. 1994; Adler and Tanner 2013), changing the environment to suit themselves in ways that impact upon the structure of the landscape – and by extension, on a range of other species as well (Hastings et al. 2007).

Ecosystem engineers are organisms that "directly or indirectly modulate the availability of resources to other species, by causing physical state changes in biotic or abiotic materials. In so doing they modify, maintain and create habitats" (Jones et al. 1994). Well-known examples of ecosystem engineers include beavers that build dams on streams, prairie dogs that dig networks of burrows, and coral polyps that aggregate to form extensive undersea reefs (Wright et al. 2002; Van Nimwegen et al. 2008; Wild et al. 2011; Bozec et al. 2013). Humans can be classified as allogenic ecosystem engineers, altering their environment by transforming materials from one physical state to another (Jones et al. 1994). In contrast, autogenic engineers, such as corals, change their environment via their own physical structures (bodies). The presence of a dominant ecosystem engineer in a non-urban ecosystem does not mean that the ecology of that ecosystem cannot be understood using existing ecological theory. Instead, the engineers are part of the theory (e.g., Jones et al. 1997). Similarly, there is no reason why the presence of humans as a dominant ecosystem engineer in cities should leave the ecology of urban environments so changed that existing ecological theories no longer apply there.

8.2.3 The uniqueness of urban ecosystems

Are urban ecosystems really unique? In some ways, yes – certain characteristics such as a high cover of impermeable surface, a high density of people and built structures, and high levels of air, noise and light pollution distinguish urban ecosystems from other ecosystems. Urban landscapes also tend to be spatially heterogeneous (Luck and Wu 2002; Cadenasso et al. 2007; Luo and Wei 2009), but this in itself is not sufficient to differentiate them from non-urban landscapes.

As discussed in Chapter 7, some urban ecosystems may represent novel ecosystems, with a combination of novel environmental conditions and novel habitat features (such as buildings, road corridors and artificial sea walls) that support new ecological communities comprising a mixture of native and introduced species (Hobbs et al. 2006; Standish et al. 2013). However, all ecosystems are unique in some way. For example, forests are distinguished from other ecosystems by a high density of trees, and a tropical rainforest is distinguished from a temperate deciduous forest by a combination of local environmental variables (rainfall, temperature, humidity, soil nutrients and moisture levels), the structural forms of the vegetation, and the identity of the plant, animal, fungal and bacterial species that inhabit each forest type.

The uniqueness of urban ecosystems does not mean that they have a unique ecology (Niemelä 1999; Catterall 2009), and the novelty of urban ecosystems does not mean that we need a novel ecological theory to understand them. Populations of all types (including human populations) in urban environments are subject to the same processes of birth, death, immigration and emigration as populations in other types of environments (Chapter 3). Ecological communities in cities are shaped by the same processes of selection, dispersal, ecological drift and evolutionary diversification that operate in deserts, grasslands, forests, streams and oceans (Chapter 4). Despite the disruptions caused by the construction and operation of cities, urban ecosystems still function to cycle carbon, water and nutrients, and to provide services such as primary production, absorption of atmospheric pollutants, shading, cooling, the biological control of pest species and the treatment of waste (Chapters 5–7). Answering the challenge of Collins et al. (2000), classical ecological theory *can* explain the distribution, abundance and relationships of organisms with their environments in cities. The presence of many human agents, their actions, preferences, motivations, power structures and socio-economic conditions all affect the biophysical characteristics of urban landscapes on a variety of spatial and temporal scales, as well as the identity, distribution and abundance of non-human species in cities. However, they do not alter the fundamental ecology of urban environments.

8.3 The definition and scope of urban ecology

If – as I have argued – existing ecological theory *is* sufficient to understand the ecology of urban environments across all levels of organization, from that of individual organisms to populations, ecological communities and urban ecosystems as a whole, then why the continuing call for a new theory of urban ecology? The answer may lie in the definition of urban ecology being used by those who argue against the applicability or adequacy of existing ecological theories. For example, Marzluff et al. (2008) described urban ecology as an emerging interdisciplinary field with deep roots in a variety of disciplines, including ecology, sociology, geography, urban planning, engineering, economics, public health and anthropology.

McDonnell (2011; 2015) proposed that urban ecology is emerging as a truly inter- and transdisciplinary science that draws upon many different natural and social sciences, while Wu (2014) declared that "... urban ecology has evolved into a truly transdisciplinary enterprise that integrates ecological, geographical, planning, and social sciences".

However, from a taxonomic perspective, an entity cannot be comprised of itself plus a range of other entities. Thus, urban ecology cannot logically include urban ecology *plus* urban planning, sociology, geography and so on and still be classified as urban ecology; this interdisciplinary agglomeration must be something broader. I propose we call it urban science, analogous to agricultural science or forest science. Agricultural science encompasses a diversity of disciplines relevant to the development, structure, function, dynamics and management of agricultural ecosystems, including agro-ecology, soil science, agronomy, genetics, animal husbandry, entomology, plant pathology, economics, cultural geography and sociology. Similarly, forest science incorporates a variety of disciplines such as forest ecology, fire ecology, soil science, silviculture, genetics, hydrology, wood science, forest planning, harvesting, sociology and economics. Urban science could be considered as a corresponding interdisciplinary field that integrates scholarship from a wide range of disciplines (including urban ecology) to study the development, dynamics, function and management of urban environments; the motivations, actions, health and wellbeing of urban-dwelling humans; and the impacts of urbanization on ecosystems and all the species that live in cities.

8.4 Do we need a new theory of urban science?

Existing ecological theory is not sufficient to understand every aspect of agricultural environments and their management (including their physical, biological, economic, political and social drivers, processes, feedbacks and interdependencies), just as existing ecological theory is not sufficient to understand all aspects of the management of forested environments for timber, recreation, conservation and other ecosystem services. Therefore, it is not surprising that existing ecological theory does not explain every aspect of constructed urban environments, from the political, social and cultural drivers of urbanization to the design, planning and management of urban systems. But does this mean we need a new theory of urban science that can explain all aspects of urbanization and urban systems – an urban theory of everything?

I argue that it does not. Inter- and transdisciplinary science conducted in any type of ecosystem can draw upon existing theory in each of the component disciplines relevant to the particular problem or question at hand. It is not necessary to develop a new, general theory to underpin every possible problem or question within the urban sphere, across all possible disciplines. For example, let us consider the following question: in a given city, how do we design, create and manage parklands to maximize their combined benefits for biodiversity, air quality, microclimatic cooling and human health and wellbeing? To address this question, we need input from a wide range of disciplines, including ecology, conservation biology, climatology, atmospheric chemistry, plant physiology, hydrology, population health, health promotion and education, environmental psychology, sociology, urban planning, economics and park management. Each discipline has its own theoretical foundations that can be used to frame the relevant part of the overall problem; surely it is more efficient (and possibly more robust) to apply this existing scholarship to interdisciplinary questions within the urban realm than to seek to reinvent the theoretical wheel. Some conceptual frameworks designed to combine data, approaches and theories from multiple disciplines already exist (e.g., Collins et al. 2011). While an integrated theory across all relevant disciplines might be a useful long-term goal within urban science, it is not required for effective planning and management of cities for biodiversity, improved ecosystem function, and/or human health and wellbeing.

8.5 Future directions

Where to from here for the discipline of urban ecology? If we accept that the ecology of urban environments can be addressed and progressed using existing ecological theories, this frees us to focus our efforts on a range of interesting urban-ecological questions, such as:

1. Can we predict the impact of the various biophysical processes of urbanization on different species within a taxonomic group based on their ecological or physiological traits?
2. How do the biophysical processes of urbanization affect population processes (births, deaths, immigration and emigration) in humans and other species?
3. What influence does the trophic position of a species have on its probability of persistence or local extinction in urban environments?
4. Does this depend on the overall trophic structure of an ecological community, pre- and post-urbanization?
5. What role does dispersal play in maintaining diverse ecological communities in cities?
6. How much habitat connectivity must we maintain to ensure adequate dispersal of organisms across the urban landscape?
7. Are certain characteristics of the urban environment driving evolutionary change within non-human species, from microbes to plants and animals?
8. By what mechanisms and to what extent do the actions of humans as ecosystem engineers in cities affect the persistence of non-human populations, the composition of ecological communities and the function of urban ecosystems?
9. Are these mechanisms of equal importance in different types of cities and different biomes around the world?

10 Can we use an improved understanding of these mechanisms to reverse (in existing urban landscapes) or prevent (in future urban landscapes) some of the undesirable impacts of urbanization on biodiversity, ecosystem functions and ecosystem services?
11 How do novel environmental stressors in cities-such as traffic noise, artificial night lighting and chemical pollution of air, water and soil-affect individuals, populations and ecological communities in terrestrial, freshwater and marine habitats?
12 To what extent can we reverse the undesirable effects of urbanization on ecological communities through habitat restoration?

There are also many intriguing questions to be addressed that fall within the broader domain of urban science, such as:

1 How can we better integrate urban ecology into urban planning and design to enhance biodiversity and ecosystem services in urban environments?
2 How can we use biodiversity and green infrastructure to improve the connectedness of urban-dwelling humans to nature and to each other, and to overcome environmental injustice caused by an inequitable distribution of ecosystem services across cities?
3 What small-scale changes can we make to improve the liveability of existing urban neighbourhoods?
4 How can we build the cities of the future to give all their human residents an adequate standard of living, while also considering the requirements of other species and maintaining valuable ecosystem functions and services?

Addressing questions such as these using the theories we already have (ecological and otherwise, as appropriate) is likely to advance the discipline of urban ecology and the interdisciplinary field of urban science more effectively than a continued search for the urban theory of everything. Given what is at stake – the persistence of biodiversity (including many threatened species) in urban environments, the structure of urban ecological communities, the function of ecosystems on a local and global scale, and the health and wellbeing of billions of people who live in cities around the world – effective advancement is paramount. We have no time to lose.

Study questions

1 Describe the characteristics of urban ecosystems that set them apart from other types of ecosystems.
2 Does the human domination of urban environments change their fundamental ecology? Discuss.
3 Do you think we need a new theory of urban ecology? Why or why not?

4 Is urban science a useful term to describe the interdisciplinary study of cities, including urban ecology, sociology, geography, urban planning and management? What other terms could be used instead, and what are their relative merits?
5 Which combination(s) of disciplinary perspectives within urban science do you think will be most important for improving the wellbeing of human and non-human populations in cities over the coming decades?

References

Adler, F.R. and Tanner, C.J. (2013) *Urban Ecosystems: Ecological Principles for the Built Environment*, Cambridge University Press, Cambridge.

Alberti, M., Marzluff, J.M., Shulenberger, E. *et al.* (2003) Integrating humans into ecology: opportunities and challenges for studying urban ecosystems. *BioScience*, **53**, 1169–1179.

Alberti, M. (2008) *Advances in Urban Ecology: Integrating Humans and Ecological Processes in Urban Ecosystems*, Springer, New York.

Anand, M., Gonzales, A., Guichard, F. *et al.* (2010) Ecological systems as complex systems: challenges for an emerging science. *Diversity*, **2**, 395–410.

Andersson, E. (2006) Urban landscapes and sustainable cities. *Ecology and Society*, **11**, 34.

Boit, A., Martinez, N.D., Williams, R.J., and Gaedke, U. (2012) Mechanistic theory and modelling of complex food-web dynamics in Lake Constance. *Ecology Letters*, **15**, 594–602.

Bozec, Y., Yakob, L., Bejarano, S. *et al.* (2013) Reciprocal facilitation and non-linearity maintain habitat engineering on coral reefs. *Oikos*, **122**, 428–440.

Cadenasso, M.L. and Pickett, S.T.A. (2008) Urban principles for ecology landscape design and management: scientific fundamentals. *Cities and the Environment*, **1**, 1–16.

Cadenasso, M.L., Pickett, S.T.A., and Schwarz, K. (2007) Spatial heterogeneity in urban ecosystems: reconceptualizing land cover and a framework for classification. *Frontiers in Ecology and the Environment*, **5**, 80–88.

Catterall, C.P. (2009) Responses of faunal assemblages to urbanisation: Global research paradigms and an avian case study, in *Ecology of Cities and Towns: A Comparative Approach* (eds M.J. McDonnell, A.K. Hahs, and J. Brueste), Cambridge University Press, Cambridge, pp. 129–155.

Collins, J.P., Kinzig, A., Grimm, N.B. *et al.* (2000) A new urban ecology: modelling human communities as integral parts of ecosystems poses special problems for the development and testing of ecological theory. *American Scientist*, **88**, 416–425.

Collins, S.L., Carpenter, S.R., Swinton, S.M. *et al.* (2011) An integrated conceptual framework for long-term social-ecological research. *Frontiers in Ecology and the Environment*, **9**, 351–357.

George, M.R., Brown, J.R., and Clawson, W.J. (1992) Application of nonequilibrium ecology to management of Mediterranean grasslands. *Journal of Range Management*, **45**, 436–440.

Hastings, A., Byers, J.E., Crooks, J.A. *et al.* (2007) Ecosystem engineering in space and time. *Ecology Letters*, **10**, 153–164.

Hobbs, R.J., Yates, S., and Mooney, H.A. (2007) Long-term data reveal complex dynamics in grassland in relation to climate and disturbance. *Ecological Monographs*, **77**, 545–568.

Hobbs, R.J., Arico, S., Aronson, J. *et al.* (2006) Novel ecosystems: theoretical and management aspects of the new ecological world order. *Global Ecology and Biogeography*, **15**, 1–7.

Jones, C.G., Lawton, J.H., and Shachak, M. (1994) Organisms as ecosystem engineers. *OIKOS*, **69**, 373–386.

Jones, C.G., Lawton, J.H., and Shachak, M. (1997) Positive and negative effects of organisms as physical ecosystem engineers. *Ecology*, **78**, 1946–1957.

Laurance, W.F., Andrade, A.S., Magrach, A. *et al.* (2014) Apparent environmental synergism drives the dynamics of Amazonian forest fragments. *Ecology*, **95**, 3018–3026.

Levin, S.A. (1998) Ecosystems and the biosphere as complex adaptive systems. *Ecosystems*, **1**, 431–436.

Liu, J., Dietz, T., Carpenter, S.R. *et al.* (2007) Complexity of coupled human and natural systems. *Science*, **317**, 1513–1516.

Luo, J. and Wei, Y.H.D. (2009) Modeling spatial variations of urban growth patterns in Chinese cities: the case of Nanjing. *Landscape and Urban Planning*, **91**, 51–64.

Luck, M. and Wu, J. (2002) A gradient analysis of urban landscape pattern: a case study from the Pheonix metropolitan region, Arizona, USA. *Landscape Ecology*, **17**, 327–339.

Marzluff, J.M., Shulenberger, E., Simon, U. *et al.* (2008) An introduction to urban ecology as an interaction between humans and nature, in *Urban Ecology: An International Perspective on the Interaction Between Humans and Nature* (eds J.M. Marzluff, E. Shulenberger, W. Endlicher, *et al.*), Springer, New York, pp. vii–xi.

McDonnell, M.J. (2011) The history of urban ecology, in *Urban Ecology: Patterns* (ed J. Niemelä), *Processes and Applications*. Oxford University Press, Oxford, pp. 5–13.

McDonnell, M.J. (2015) Journal of Urban Ecology: Linking and promoting research and practice in the evolving discipline of urban ecology. *Journal of Urban Ecology*, **1**, 1–6.

McDonnell, M.J. and Pickett, S.T.A (eds). (1997) *Humans as Components of Ecosystems: The Ecology of Subtle Human Effects and Populated Areas*. Springer-Verlag, New York.

Niemelä, J. (1999) Is there a need for a theory of urban ecology? *Urban Ecosystems*, **3**, 57–65.

Pen-Mouratov, S. and Steinberger, Y. (2005) Spatio-temporal dynamic heterogeneity of nematode abundance in a desert ecosystem. *Journal of Nematology*, **37**, 26–36.

Pickett, S.T.A., Cadenasso, M.L., Grove, J.M. *et al.* (2008) Beyond urban legends: an emerging framework of urban ecology, as illustrated by the Baltimore Ecosystem Study. *BioScience*, **58**, 139–150.

Qureshi, S., Haase, D., and Coles, R. (2014) The Theorized Urban Gradient (TUG) method – a conceptual framework for socio-ecological sampling in complex urban agglomerations. *Ecological Indicators*, **36**, 100–110.

Schuur, E.A.G., McGuire, A.D., Schadel, C. *et al.* (2015) Climate change and the permafrost carbon feedback. *Nature*, **520**, 171–179.

Standish, R.J., Hobbs, R.J., and Miller, J.R. (2013) Improving city life: options for ecological restoration in urban landscapes and how these might influence interactions between people and nature. *Landscape Ecology*, **28**, 1213–1221.

Sugihara, G., May, R., Ye, H. *et al.* (2012) Detecting causality in complex ecosystems. *Science*, **338**, 496–500.

Van Nimwegen, R.E., Kretzer, J., and Cully, J.F. Jr., (2008) Ecosystem engineering by a colonial mammal: how prairie dogs structure rodent communities. *Ecology*, **89**, 3298–3305.

Vitousek, P.M., Mooney, H.A., Lubchenco, J. *et al.* (1997) Human domination of Earth's ecosystems. *Science*, **277**, 494–499.

Wild, C., Hoegh-Guldberg, O., Naumann, M.S. *et al.* (2011) Climate change impedes scleractinian corals as primary reef ecosystem engineers. *Marine and Freshwater Research*, **62**, 205–215.

Wipf, S., Stoeckli, V., and Bebi, P. (2009) Winter climate change in alpine tundra: plant responses to changes in snow depth and snowmelt timing. *Climate Change*, **94**, 105–121.

Wright, J.P., Jones, C.G., and Flecker, A.S. (2002) An ecosystem engineer, the beaver, increases species richness at the landscape scale. *Oecologia*, **132**, 96–101.

Wu, J. (2014) Urban ecology and sustainability: the state-of-the-science and future directions. *Landscape and Urban Planning*, **125**, 209–221.

Index

Ecology of Urban Environments, First Edition. Kirsten M. Parris.

CPSIA information can be obtained
at www.ICGtesting.com
Printed in the USA
JSHW050730250723
45340JS00010B/161

9 781444 332650